# BASIC DIGITAL ELECTRONICS—

## Understanding Number Systems, Boolean Algebra, & Logic Circuits

# BASIC DIGITAL ELECTRONICS-

## Understanding Number Systems, Boolean Algebra, & Logic Circuits

### by Ray Ryan

TAB BOOKS Inc.
BLUE RIDGE SUMMIT, PA. 17214

FIRST EDITION

EIGHTH PRINTING

Printed in the United States of America

ISBN 0-8306-4728-7
ISBN 0-8306-3728-1 (pbk.)

Library of Congress Card Number: 74-14326

# Introduction

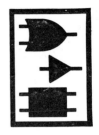

Digital circuits are finding application in every phase of electronics today. They are found in such diverse applications as TV receivers, stereo circuits, telephone circuits, and even in automobiles. Even in the rf and high power areas, digital circuits are performing an increasingly large portion of the display and control functions.

This book is intended as a reference book for people who are already familiar with basic electronics, but to whom digital circuits are still an unexplored mystery. It is hoped that this book will introduce the ones and zeroes of the digital world in a simplified and straightforward manner so that any technician, engineer, or hobbyist can understand the principles involved without having to resort to higher mathematics or special training.

The main emphasis throughout the book is on presenting the various types of circuits and describing the features pertinent to these circuits. Today, discrete circuits are seldom used, and almost all logic is performed with integrated circuits. For this reason, there is a slight bias throughout toward the assumption that a given logic function is just that, a *function*, instead of being concerned with specific circuit details.

Because a system of ones and zeroes probably seems quite strange to the uninitiated, the first several chapters describe the binary number system and the different kinds of codes that digital systems use.

The next several chapters explain the basics of logic and introduce the logic diagram and truth table. With this background, any digital circuit can be analyzed and understood.

The remaining chapters are basically descriptions of applications of digital circuits. Starting with very simple functions made up with AND and OR gates, the circuits become increasingly more sophisticated until, finally, memories are described that consist of literally thousands of gates and flip-flops.

It is hoped that this book fulfills the need for a simple and easily understood guide to digital circuits.

# Contents

AND GATE
FLIP-FLOP
ONE-SHOT
FREE-RUNNING MULTIVIBRATOR
SCHMITT TRIGGER
INTEGRATED CIRCUITS

RELAY LOGIC
DIODE LOGIC
DIODE—TRANSISTOR LOGIC
DIRECT COUPLED TRANSISTOR LOGIC
RESISTOR—TRANSISTOR LOGIC
RESISTOR—CAPACITOR—TRANSISTOR LOGIC
TRANSISTOR—TRANSISTOR LOGIC
EMITTER-COUPLED LOGIC
MOS LOGIC

FLIP-FLOPS
    NAND Gate SR Flip-Flop
    NOR Gate SR Flip-Flop
    Clocked Flip-FLops
    Clocked SR Flip-Flop
    T Flip-Flop
    D Flip.Flop
    JK Flip-Flop
MOS DYNAMIC STORAGE

ALGEBRAIC MANIPULATION
KARNAUGH MAPS
FLOW DIAGRAMS
RACE CONDITIONS

REGISTERS
    Delay Line
    Storage Register
    Serial-to-Parallel Converter
    Parallel-to-Serial Converter
COUNTERS
    Ripple Counter
    Synchronous Counter
    BCD Counter
    Johnson Counter
    Ring Counter
DECODERS
    Single Decode Gate
    Don't-Care Conditions
    Decoders for Johnson and Ring Counters
ADDERS
    Serial Adder
    Parallel Adder

# List of Illustrations
# and
# List of Tables

# Number Systems

A digital circuit is a circuit that expresses voltage or current values as digits. Usually, there are only two digits, 1 and 0. The 1 can be thought of as representing a transistor or vacuum tube that is forward biased or a relay contact that is closed. The 0 then indicates a cut-off transistor or tube, or an open relay contact. This is not a complex idea. In fact, most digital circuits are quite simple and easy to understand.

The numbering concept that uses only the digits 1 and 0 is called a *binary* system. As will be seen, the binary system works very much like the familiar decimal (having ten digits) system. By knowing a few basic rules governing number systems, one of the most important uses for digital circuits, that of performing arithmetic, can be understood. It is the purpose of this chapter to introduce three basic number systems—decimal, binary, and octal—and to outline the rules for using them.

## DECIMAL NUMBER SYSTEM

The most common number system, and the one with which everyone is familiar, is the decimal number system. Decimal numbers are represented with these digits: 0, 1, 2, 3, 4, 5, 6, 7, 8, and 9. Each position in a decimal number has a value that is 10 times the value of the next position to the right. In other words, every position can be expressed by 10 raised to some power. The *unit* or "ones" position is $10^0$ (since any number raised to the zero power, except zero, is equal to one), the *tens* position is $10^1$, and the *hundreds* position is $10^2$. A progression of increasing exponents (superscript numerals) can be continued as far as desired to the left of the decimal point. The same progression can also be extended to the right of the decimal point, but the exponents will be negative. For example, the first position to the right of the decimal point is the *tenths* position; it has a weight of $10^{-1}$. Thus, the form of every decimal number is:

$$\ldots 10^4\, 10^3\, 10^2\, 10^1\, 10^0 \,.\, 10^{-1}\, 10^{-2}\, 10^{-3} \ldots$$

Note that in the decimal system the most significant digit (the one with the largest power of 10) is always written at the left, and the least significant digit is written at the right. This convention will be used throughout the book when writing numbers in any number system.

Other number systems can be represented in a fashion similar to the decimal system. In each case, the weights of the different positions are powers of a particular number representing the quantity of digits available in that number system. This number is called the *radix* of the number system. Using the letter **R** to represent the radix of any number system, the form of any number is:

$$... R^4 R^3 R^2 R^1 R^0 . R^{-1} R^{-2} R^{-3} ...$$

If there is doubt as to the number system being employed, it should be clarified by writing the radix of the number system as a subscript to the number. For example, $235_{10}$ indicates that the number 235 is written in the decimal number system.

## BINARY NUMBER SYSTEM

The binary number system is the system most useful in digital circuits, because in this system there are only two digits (0 and 1) and these two digits may be used to represent the states (either closed or open; either on or off) of switches, vacuum tubes, relays, transistors, etc. Since the binary number system uses two digits, it has a radix of two, and the position weights are powers of two. The progression of powers in binary is:

$$...2^0 = 1 \quad 2^1 = 2 \quad 2^2 = 4 \quad 2^3 = 8 \quad 2^4 = 16 ...$$

A list of powers of two is provided in Table 1-1, for reference when dealing with binary numbers. Counting in the binary number system is performed much the same as in the decimal number system. Recall that with decimal numbers the count proceeds: 0, 1,...8, 9, until a single digit can no longer represent the number. Then, a second digit is added and the count continues; 10, 11,...20,...30...etc. This adding of digits is continued for higher and higher counts so that there are always enough digits to represent any number. The binary counts progress in a similar manner. Counting from zero to four in binary, the counts are: 0, 1, 10, 11, 100. Note that each time the two digits in one position are exhausted, a 1 is added at the left, all other digits are made 0, and the count is continued. Of course, there is no reason that decimal numbers could not include leading zeroes, just so

Table 1-1. Powers of Two.

| $n$ | $2^n$ | $n$ | $2^n$ |
|---|---|---|---|
| 1 | 2 | 41 | 21990 23255 552 |
| 2 | 4 | 42 | 43980 46511 104 |
| 3 | 8 | 43 | 87960 93022 208 |
| 4 | 16 | 44 | 17592 18604 4416 |
| 5 | 32 | 45 | 35184 37208 8832 |
| 6 | 64 | 46 | 70368 74417 7664 |
| 7 | 128 | 47 | 14073 74883 55328 |
| 8 | 256 | 48 | 28147 49767 10656 |
| 9 | 512 | 49 | 56294 99534 21312 |
| 10 | 1024 | 50 | 11258 99906 84262 4 |
| 11 | 2048 | 51 | 22517 99813 68524 8 |
| 12 | 4096 | 52 | 45035 99627 37049 6 |
| 13 | 8192 | 53 | 90071 99254 74099 2 |
| 14 | 16384 | 54 | 18014 39850 94819 84 |
| 15 | 32768 | 55 | 36028 79701 89639 68 |
| 16 | 65536 | 56 | 72057 59403 79279 36 |
| 17 | 13107 2 | 57 | 14411 51880 75855 872 |
| 18 | 26214 4 | 58 | 28823 03761 51711 744 |
| 19 | 52428 8 | 59 | 57646 07523 03423 488 |
| 20 | 10485 76 | 60 | 11529 21504 60684 6976 |
| 21 | 20971 52 | 61 | 23058 43009 21369 3952 |
| 22 | 41943 04 | 62 | 46116 86018 42738 7904 |
| 23 | 83886 08 | 63 | 92233 72036 85477 5808 |
| 24 | 16777 216 | 64 | 18446 74407 37095 51616 |
| 25 | 33554 432 | 65 | 36893 48814 74191 03232 |
| 26 | 67108 864 | 66 | 73786 97629 48382 06464 |
| 27 | 13421 7728 | 67 | 14757 39525 89676 41292 8 |
| 28 | 26843 5456 | 68 | 29514 79051 79352 82585 6 |
| 29 | 53687 0912 | 69 | 59029 58103 58705 65171 2 |
| 30 | 10737 41824 | 70 | 11805 91620 71741 13034 24 |
| 31 | 21474 83648 | 71 | 23611 83241 43482 26068 48 |
| 32 | 42949 67296 | 72 | 47223 66482 86964 52136 96 |
| 33 | 85899 34592 | 73 | 94447 32965 73929 04273 92 |
| 34 | 17179 86918 4 | 74 | 18889 46593 14785 80854 784 |
| 35 | 34359 73836 8 | 75 | 37778 93186 29571 61709 568 |
| 36 | 68719 47673 6 | 76 | 75557 86372 59143 23419 136 |
| 37 | 13743 89534 72 | 77 | 15111 57274 51828 64683 8272 |
| 38 | 27487 79069 44 | 78 | 30223 14549 03657 29367 6544 |
| 39 | 54975 58138 88 | 79 | 60446 29098 07314 58735 3088 |
| 40 | 10995 11627 776 | 80 | 12089 25819 61462 91747 06176 |

| $n$ | $2^n$ |
|---|---|
| 81 | 24178 51639 22925 83494 12352 |
| 82 | 48357 03278 45851 66988 24704 |
| 83 | 96714 06556 91703 33976 49408 |
| 84 | 19342 81311 38340 66795 29881 6 |
| 85 | 38685 62622 76681 33590 59763 2 |
| 86 | 77371 25245 53362 67181 19526 4 |
| 87 | 15474 25049 10672 53436 23905 28 |
| 88 | 30948 50098 21345 06872 47810 56 |
| 89 | 61897 00196 42690 13744 95621 12 |
| 90 | 12379 40039 28538 02748 99124 224 |
| 91 | 24758 80078 57076 05497 98248 448 |

that they would all appear to have the same number of digits. For example, if 999 were the highest number to be represented, then the lower decimal counts could be written 000, 001, 002, 003,...etc. A typical binary count sequence using four binary digits is given in Table 1-2. The leading zeroes can be disregarded, if desired, but they are normally included for symmetry and convenience.

Table 1-2. Four Digit Binary Count Sequence.

| DECIMAL NUMBER | BINARY NUMBER |
|:---:|:---:|
| 0 | 0000 |
| 1 | 0001 |
| 2 | 0010 |
| 3 | 0011 |
| 4 | 0100 |
| 5 | 0101 |
| 6 | 0110 |
| 7 | 0111 |
| 8 | 1000 |
| 9 | 1001 |
| 10 | 1010 |
| 11 | 1011 |
| 12 | 1100 |
| 13 | 1101 |
| 14 | 1110 |
| 15 | 1111 |

**Binary-to-Decimal Conversion**

Binary numbers can be converted to decimal form quite simply. The method for performing this conversion consists of adding together all of the position weights where a one appears. All positions containing a zero can be ignored. The entire procedure can be thought of as a two-step process. First, the position weights are determined by referring to Table 1-1 and the proper position weights are written above the binary number which is to be converted. Second, the position weights containing one's are added to obtain the resultant decimal number. This simple procedure works equally well for converting whole numbers or fractional numbers. Examples of several binary-to-decimal conversions:

## Binary-to-decimal conversion

<u>32 16 8 4 2 1</u>
$$1\ 1\ 0\ 1\ 0\ 1 = 32 + 16 + 4 + 1 = 53$$
$$1\ 0\ 1\ 0\ 0\ 1 = 32 + 8 + 1 = 41$$

## Decimal-to-Binary Conversion

Decimal to binary conversion is a somewhat more lengthy process. Also, there are several commonly used techniques for performing the conversion. The first method used is the repeated subtraction of powers of two until either there is no remainder, or until the remainder which is left is sufficiently small for the desired accuracy of conversion. This method works best for whole numbers primarily because the positive powers of two are easily remembered and also because there are not too many digits to deal with. A second method consists of breaking the decimal number into two parts, (the whole number part and the fractional number part), and separately converting each part. The whole number part can be converted by successively dividing the number by two, each time noting whether or not the number is exactly divisible by two. Each time there is a remainder of zero, a zero is written down and each time there is a remainder of one, a one is written down. The fractional part of the decimal number is converted by repeatedly multiplying by two until the decimal portion of the number becomes exactly zero, or until sufficient accuracy is obtained. This method works well for fractional numbers because it is not necessary to know the negative powers of two to perform the conversion. Because the two methods differ significantly, each of the methods will be described separately.

*Subtraction Method.* To perform a decimal-to-binary conversion by the subtraction mehod, simply write down the powers of two up to and including the one closest to the number being converted and begin to sequentially subtract each of these powers of two from the decimal number. If the power can be subtracted—that is, if it's not larger than the decimal remainder—then a binary 1 is written in the bit position corresponding to that power of two. If the power is larger than the decimal remainder, a 0 is written in that position and the next lower power of two is attempted. For example, consider the conversion for the decimal number 50. The largest power of two which is still smaller than 50 is $2^5 = 32$. The powers of two from $2^5$

down are written and the resultant subtractions are shown below.

Decimal-to-binary conversion (subtraction method)

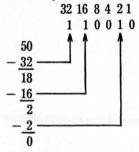

**Two-Part Method.** To convert the whole number part by this method, the decimal number is written down, then repeatedly divided by two. Each time there is no "remainder" a 0 is written down for the binary number. When there is a remainder, a one is written down for the binary number. The first division produces the least significant digit and the last division produces the most significant digit. A conversion for the decimal number 53 is shown below. Note that this method does not require writing down the powers of two.

Decimal-to-binary conversion (whole part — division)

| Division | Remainder | |
|----------|-----------|--|
| $53 \div 2 = 26$ | 1 | least significant digit |
| $26 \div 2 = 13$ | 0 | |
| $13 \div 2 = 6$ | 1 | |
| $6 \div 2 = 3$ | 0 | |
| $3 \div 2 = 1$ | 1 | |
| $1 \div 2 = 0$ | 1 | most significant digit |

Result: $53_{10} = 1\ 1\ 0\ 1\ 0\ 1_2$

The fractional number part is converted by repeatedly multiplying by two. If the multiplication produces a whole number equal to one plus some fractional part, a one is written down for the binary number and the whole number part of the result is dropped. If the multiplication produces a whole number equal to zero plus a fractional part, a zero is written down. The multiplications by two are repeated until the fractional part exactly equals zero, or until a sufficient number of digits have been obtained. An example of the conversion of the decimal number 0.3125 is shown next. As before, note that it is not necessary to know the powers of two to perform the conversion.

## Decimal-to-binary conversion (fractional part—multiplication)

| Multiplication | Whole Number Part | |
|---|---|---|
| .3125 | | |
| 2 | | |
| 0.6250 | 0 | most significant digit |
| 2 | | |
| 1.2500 | 1 | |
| 2 | | |
| 0.5000 | 0 | |
| 2 | | |
| 1.0000 | 1 | least significant digit |

Result: $0.3125_{10} = 0 . 0 1 0 1_2$

**Signed Binary Numbers**

Now that the concept of a simple binary number is understood, the representation of positive and negative numbers will be considered. Binary numbers which carry identification as to their polarity are referred to as signed binary numbers. Of practical interest is a method by which the plus and minus signs for positive and negative numbers can be carried around in a digital format. Further, if a binary number is negative, it turns out that there are several convenient ways of representing that number. Each representation has its own unique features. The three major signed binary number notations are (1) *sign magnitude* notation, (2) *one's complement* notation, and (3) *two's complement* notation.

The most straightforward way to identify a signed binary number is to simply add a 0 or a 1 to the most significant bit (bit is short for binary digit) of the overall number. This notation is called **sign magnitude** notation because the sign bit is given first, then the positive magnitude of the number is given next. A sign bit of zero indicates the number is positive, while a negative number is indicated by a sign bit of one. All other bits of the number indicate the magnitude of the number just as for an unsigned binary number. Sign magnitude notation is very easy to read, but as will be seen, it is not easy to use for addition and subtraction. An example of the decimal number 13 shown as both a positive and a negative sign magnitude number is given next. Also, Table 1-3 shows the digits from −15 to +15 in sign

**Table 1-3. Comparison of Signed Binary Numbers.**

| DECIMAL | SIGN MAGNITUDE | ONE'S COMPLEMENT | TWO'S COMPLEMENT |
|---------|----------------|------------------|------------------|
| 15 | 0 1111 | 0 1111 | 0 1111 |
| 14 | 0 1110 | 0 1110 | 0 1110 |
| 13 | 0 1101 | 0 1101 | 0 1101 |
| 12 | 0 1100 | 0 1100 | 0 1100 |
| 11 | 0 1011 | 0 1011 | 0 1011 |
| 10 | 0 1010 | 0 1010 | 0 1010 |
| 9 | 0 1001 | 0 1001 | 0 1001 |
| 8 | 0 1000 | 0 1000 | 0 1000 |
| 7 | 0 0111 | 0 0111 | 0 0111 |
| 6 | 0 0110 | 0 0110 | 0 0110 |
| 5 | 0 0101 | 0 0101 | 0 0101 |
| 4 | 0 0100 | 0 0100 | 0 0100 |
| 3 | 0 0011 | 0 0011 | 0 0011 |
| 2 | 0 0010 | 0 0010 | 0 0010 |
| 1 | 0 0001 | 0 0001 | 0 0001 |
| +0 | 0 0000 | 0 0000 | 0 0000 |
| −0 | | 1 1111 | |
| −1 | 1 0001 | 1 1110 | 1 1111 |
| −2 | 1 0010 | 1 1101 | 1 1110 |
| −3 | 1 0011 | 1 1100 | 1 1101 |
| −4 | 1 0100 | 1 1011 | 1 1100 |
| −5 | 1 0101 | 1 1010 | 1 1011 |
| −6 | 1 0110 | 1 1001 | 1 1010 |
| −7 | 1 0111 | 1 1000 | 1 1001 |
| −8 | 1 1000 | 1 0111 | 1 1000 |
| −9 | 1 1001 | 1 0110 | 1 0111 |
| −10 | 1 1010 | 1 0101 | 1 0110 |
| −11 | 1 1011 | 1 0100 | 1 0101 |
| −12 | 1 1100 | 1 0011 | 1 0100 |
| −13 | 1 1101 | 1 0010 | 1 0011 |
| −14 | 1 1110 | 1 0001 | 1 0010 |
| −15 | 1 1111 | 1 0000 | 1 0001 |

magnitude notation and compares them to the two other signed binary number types.

$$\begin{aligned} \text{SIGN} \\ 0 \ \ 1 \ 1 \ 0 \ 1 &= +13 \\ 1 \ \ 1 \ 1 \ 0 \ 1 &= -13 \end{aligned}$$

Another simple way to show a negative number is to attach a sign bit as with the sign magnitude notation and invert all of the bits if the number is negative. This number representation is called *one's complement* notation. One's complement numbers are easy to form (simply invert all the bits), but as can be seen from Table 1-3, there are two representations for zero. Also, when one's complement numbers are added and subtracted, a process called end-around carry is necessary to obtain a correct answer. An example of the decimal number 13 is shown below to indicate the difference between positive and negative numbers.

SIGN
0  1 1 0 1 = + 13
1  0 0 1 0 = − 13

The most common representation for signed binary numbers is *two's complement* notation. The two's complement is generated by inverting the bits as with one's complement, then adding one to the least significant bit (LSB). Two's complement is more difficult to generate than one's complement but it simplifies addition and subtraction. Also, there is only one representation for zero in two's complement notation. The decimal number 13 is shown below in two's complement notation, to show the difference between positive and negative numbers. Refer to Table 1-3 for a comparison of two's complement numbers to the other signed binary number notations.

SIGN
0  1 1 0 1 = + 13
1  0 0 1 1 = − 13

## OCTAL NUMBER SYSTEM

The octal number system has a radix of eight and uses eight digits (0, 1, 2, 3, 4, 5, 6, 7). The position weights in the octal number system are powers of eight. The powers of eight also happen to be every third power of two. Thus, Table 1-1 can also be used as a powers-of-eight table. For reference, the first six powers of eight are:

$$8^0 = 1$$
$$8^1 = 8$$
$$8^2 = 64$$
$$8^3 = 512$$
$$8^4 = 4096$$
$$8^5 = 32768$$

The octal number system is frequently used by people involved with digital circuits, since it can easily be converted to binary. Also, there are significantly fewer digits in any given octal number than in the corresponding binary number and for this reason, it is much easier to work with octal numbers than it is to use binary numbers consisting of long strings of ones and zeroes.

Table 1-4. Relationship of Octal to Binary.

| OCTAL DIGIT | BINARY BITS |
|:-----------:|:-----------:|
| 0 | 000 |
| 1 | 001 |
| 2 | 010 |
| 3 | 011 |
| 4 | 100 |
| 5 | 101 |
| 6 | 110 |
| 7 | 111 |

## Binary-to-Octal and Octal-to-Binary Conversions

The conversion of a binary number to octal or an octal number to binary is quite simple, due to the fact that eight is the third power of two. This provides a direct correlation between 3-bit groups in a binary number and the octal digits. That is, each 3-bit group of binary bits can be represented by one octal digit. The relationship of octal digits to binary bits is given in Table 1-4. Several examples of conversions from octal to binary and binary to octal are given below.

Octal-to-binary conversion

$$
\begin{array}{c|c|c}
6 & 3 & 4 \\
110 & 011 & 100
\end{array}
$$

$$
\begin{array}{c|c|c}
2 & 1 & 7 \\
010 & 001 & 111
\end{array}
$$

Binary-to-octal conversion

$$
\begin{array}{c|c|c}
001 & 011 & 011 \\
1 & 3 & 3
\end{array}
$$

$$
\begin{array}{c|c|c}
111 & 110 & 101 \\
7 & 6 & 5
\end{array}
$$

## Decimal-to-Octal Conversion

If it is desired to convert a decimal number to octal, the simplest method is by division. The technique used is very much like the method for decimal-to-binary conversion. Begin by writing down the number to be converted, then repeatedly divide by eight. At each step of the division, any remainder represents the octal digit for one particular weight position. An example of the conversion of decimal 91 to octal:

## Decimal-to-octal conversion

| Division | Remainder | |
|---|---|---|
| $91 \div 8 = 11$ | 3 | least significant digit |
| $11 \div 8 = 1$ | 3 | |
| $1 \div 8 = 0$ | 1 | most significant digit |

Result: $91_{10} = 133_8 = 1011011_2$

One very nice advantage to the octal conversion technique is that it automatically provides a free conversion to binary. Thus, another simple way to do a decimal-to-binary conversion is to convert the decimal number to octal, then just write down the binary number from the octal. To realize the simplicity of this technique, it is only necessary to note that three divisions were required for the octal conversion of decimal 91, whereas seven divisions would be required to perform a binary conversion.

## Octal-to-Decimal Conversion

An octal-to-decimal conversion can be done in the same manner as a binary-to-decimal conversion. That is, simply add up the position weights to obtain the decimal number. Hence, to reconvert $133_8$ back to $91_{10}$, the procedure is as shown below.

### Octal-to-decimal conversion

| | | | |
|---|---|---|---|
| 64 8 1 | | | Position weights |
| 1 3 3 | | | Octal number |

$$64 \times 1 + 8 \times 3 + 1 \times 3 = 64 + 24 + 3 = 91$$

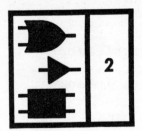

# 2 Binary Codes

Numbers expressed in signed binary fashion as described in Chapter 1 are excellent for computational purposes in the inner workings of a digital machine. However, when numbers are to be displayed, a somewhat different format is required so that operators of the machine need not deal in the awkward binary number system. Furthermore, special codes are useful when converting analog quantities to digital form and also for error detection and correction. Hence, a number of special-purpose codes have been developed, each suited to specific functional requirements. As might be surmised, there is an almost unlimited number of variations on codes which can be formed for one purpose or another. This chapter will deal only with the most common of these codes and will indicate some typical applications for each.

## BINARY-CODED DECIMAL FORMAT

As was previously indicated, circuits and machines can deal readily with binary numbers, but people are accustomed to working with decimals. Also, as has been noted, there are considerably fewer decimal digits required to represent a number than the binary ones and zeroes required to represent the same number. It is much easier to remember just a few digits than it is to remember many. Thus, whenever there is an interface between digital circuits and people, the interface data will usually take the decimal form. This requires that the digital circuits utilize some binary code to conveniently represent the decimal numbers. Several such codes are commonly used. These all come under the classification of **binary-coded decimal** (BCD) format. Typically, a binary-coded decimal contains four binary bits for each decimal digit.

### 8421 Code

One example of a binary-coded decimal format is the 8421 code. With this code, the decimal number 479 would be represented with twelve bits as 0100 0111 1001. Although this number contains only ones and zeroes, it is not a true binary number, since it does not follow the rule that each bit is weighted by an increasing power of two. Instead, each 4-bit grouping has a binary weighting, but the individual groups are powers of ten. The coding for the 10 decimal digits is shown in Table 2-1. For each 4-bit combination, the least significant bit has a weight of one, and next bit has a weight of two, the next a weight of four, and the most significant bit a weight of eight, thereby deriving its name. As can be seen, the 8421 code has a standard binary weighting for each 4 bits; therefore, it is the most common BCD code and is often referred to as simply BCD.

Table 2-1. Comparisons of BCD Codes.

| BINARY | 8421 CODE | EXCESS THREE CODE |
|--------|-----------|-------------------|
| 0000 | 0000 = 0 | |
| 0001 | 0001 = 1 | |
| 0010 | 0010 = 2 | |
| 0011 | 0011 = 3 | 0011 = 0 |
| 0100 | 0100 = 4 | 0100 = 1 |
| 0101 | 0101 = 5 | 0101 = 2 |
| 0110 | 0110 = 6 | 0110 = 3 |
| 0111 | 0111 = 7 | 0111 = 4 |
| 1000 | 1000 = 8 | 1000 = 5 |
| 1001 | 1001 = 9 | 1001 = 6 |
| 1010 | | 1010 = 7 |
| 1011 | | 1011 = 8 |
| 1100 | | 1100 = 9 |
| 1101 | | |
| 1110 | | |
| 1111 | | |

In general, BCD codes are less efficient than straight binary numbers, since they need more bits to represent a given number. For example, the number 479 required twelve bits in BCD, but can be represented with ten bits in pure binary. To be more specific, four binary bits can always represent sixteen combinations, and since there are only ten decimal digits for each four bits, a BCD code has six unused or wasted combinations.

**Excess-Three Code**

Whereas the 8421 code makes use of the first ten combinations of four bits, the **excess-three** code is a BCD format obtained by adding three to each decimal digit and coding the result as an ordinary binary number of four bits. A comparison of the 8421 code and excess-three code is provided in Table 2-1. As can be seen from the table, the excess-three code is symmetrically positioned within the 16 combination so that there are three unused states preceding and following the allowed combinations. This symmetry permits the excess-three code to be self-complementing; as a result, when arithmetic operations are described later, it will be seen that subtraction is simplified. Also, this code does not use the all-zero state; and this feature can be used as an error-detecting device. If any excess-three number should occur with no ones in it, then it would be apparent that some equipment malfunction was causing an error. As with the 8421 code, it should be noted that there are exactly six unused combinations of the four binary bits.

## Hexadecimal Code

It seems a shame to waste all those unused states for a BCD code, yet it is very nice to be able to segment the binary information into 4-bit groupings and thereby lessen the number of digits to be remembered. A code which utilizes all 16 states for each 4 bits is the hexadecimal code. This code is used frequently by digital computer personnel where very large groupings of 8, 16, or 32 bits are the normal form of data transfer. The hexadecimal code for the 16 combinations is shown in Table 2-2. From the table, it is seen that there is one digit or letter which represents each 4-bit combination. Consider a 16-bit piece of information which a computer technician might wish to analyze. The 16 bits for this example will be read in pure binary as 1110011010100101. But coded into hexadecimal, the bits are regrouped and read as shown below. Clearly, E6A5 is much easier to remember than all those ones and zeroes.

| 1110 | 0110 | 1010 | 0101 | Binary |
|------|------|------|------|--------|
| E | 6 | A | 5 | Hexadecimal |

## CYCLIC CODES

Another class of binary format is the cyclic coding system. These codes have the property of a single bit change when counting from one state to the next. Normally, a standard binary code may change several bits when going from one count to the next adjacent count. For example, when a binary number changes from 7(0111) to 8(1000), there are four bits changing simultaneously. A cyclic code would only change one of the bits, say from 0111 for 7 to 1111 for 8. Certainly this code does not have a binary weighting. The property of a single-bit change is advantageous where the output of some analog device must be

| BINARY | HEXADECIMAL CODE |
|--------|------------------|
| 0000 | 0 |
| 0001 | 1 |
| 0010 | 2 |
| 0011 | 3 |
| 0100 | 4 |
| 0101 | 5 |
| 0110 | 6 |
| 0111 | 7 |
| 1000 | 8 |
| 1001 | 9 |
| 1010 | A |
| 1011 | B |
| 1100 | C |
| 1101 | D |
| 1110 | E |
| 1111 | F |

Table 2-2. Hexadecimal Code.

digitized. Usually, an electromechanical device such as a shaft encoder, described later in this book, is used to perform the conversion. However, a shaft can stop at any random position—say, for example, on the border of changing from 0111 to 1000. If this situation occurs, each bit could be in any possible state, depending on the precision with which the electromechanical device was constructed. In this situation, the code being read out of the device is totally ambiguous. To eliminate this problem, a **cyclic** code is used. Since the code only changes one bit from position to position, the ambiguity at any position is at most one bit. The disadvantage to all cyclic codes is that arithmetic is extremely difficult to perform. If arithmetic manipulation is required, the cyclic code is usually converted to binary.

### Gray Code

The most common cyclic format is called the **gray code**. Gray code is formed from pure binary by comparing each bit with its next adjacent bit, starting with the least significant bit and forming a **modulo two** sum from the comparison. A modulo two sum is simply a binary sum with no carry. Thus, when comparing two bits, if they are alike (both ones or both zeroes) a gray code bit of zero will be generated. If the two bits are different, a gray code bit of one will be generated. For example,

$$57_{10} = \begin{array}{c} 1\ 1\ 1\ 0\ 0\ 1 \\ 0\ 1\ 1\ 1\ 0\ 0 \\ 1\ 0\ 0\ 1\ 0\ 1 \end{array} \begin{array}{l} \text{Binary} \\ \text{Shifted right one place} \\ \text{Modulo two sum} = \text{gray code} \end{array}$$

If desired, gray code can be converted back to binary by a similar process, but starting with the most significant bit. If a one occurs, the next binary bit will be the opposite of the previous bit, but if a zero occurs, the next binary bit will be the same. In the example below, the first one encountered is copied directly as it represents a change from the zero state, to the one state. Thereafter, each one represents an additional change in state from the bit just written down and each zero represents no change in state.

|  | No | No |  | No |  |  |
|---|---|---|---|---|---|---|
| Copy | Change | Change | Change | Change | Change |  |
| 1 | 0 | 0 | 1 | 0 | 1 | Gray code |
| 1 → | 1 → | 1 → | 0 → | 0 → | 1 | Binary |

A complete listing of the gray code equivalents for the first 16 binary combinations is shown in Table 2-3. Note that although several bits of the binary code may change between adjacent counts, only one bit changes for the gray code.

Table 2-3. Comparison of Gray Code to Binary Counts.

| BINARY | GRAY CODE |
|---|---|
| 0000 | 0000 |
| 0001 | 0001 |
| 0011 | 0010 |
| 0100 | 0110 |
| 0101 | 0111 |
| 0110 | 0101 |
| 0111 | 0100 |
| 1000 | 1100 |
| 1001 | 1101 |
| 1010 | 1111 |
| 1011 | 1110 |
| 1100 | 1010 |
| 1101 | 1011 |
| 1110 | 1001 |
| 1111 | 1000 |

## ERROR DETECTING CODES

Modern digital equipment is quite complex and there is always a possibility of mechanical or electrical failure in some portion of the unit. For this reason, error detecting codes have been devised which can check the result of a data transmission or an arithmetic operation. These codes consist of information bits plus some **check** bits. The check bits can be considered redundant, in that the resultant code contains more bits than are acutally necessary to represent a particular piece of information.

The BCD codes previously described already contain is limited amount of error detecting capability. Since BCD uses only 10 of 16 states for each 4-bit group, the 6 states which do not occur represent "forbidden" combinations. If any of these combinations does occur, it is as the result of an error. Also, as was previously mentioned, the excess-three code has the additional error-detecting feature that the all-zero state is one of the forbidden combinations. Since an all-zero state represents the absence of a signal, this is a useful fault-detecting feature in digital circuits.

The most common method of error detecting is through the use of **parity bits**. Usually, a single parity bit is added to the information bits, such that the number of ones in the data is odd or even, as desired. For example, suppose that the binary number 1001 were to be transmitted with a parity bit added which would make the number of ones odd. Then, the actual binary data would consist of five bits, as shown below.

$$1\ 0\ 0\ 1 \qquad\qquad 1$$
data bits    parity bits

This scheme is called **odd parity**, because it is always desired that there be an odd number of ones in the data. The proper odd parity bits for the first 16 binary combinations are shown in Table 2-4. In a similar fashion, **even parity** can also be used. Here, the object is to always have an even number of ones in the data. The parity bit for an even parity scheme is the opposite of that required to form odd parity.

Table 2-4. Odd Parity for Four Data Bits.

| DECIMAL NUMBER | BINARY NUMBER | PARITY BIT |
|---|---|---|
| 0 | 0000 | 1 |
| 1 | 0001 | 0 |
| 2 | 0010 | 0 |
| 3 | 0011 | 1 |
| 4 | 0100 | 0 |
| 5 | 0101 | 1 |
| 6 | 0110 | 1 |
| 7 | 0111 | 0 |
| 8 | 1000 | 0 |
| 9 | 1001 | 1 |
| 10 | 1010 | 1 |
| 11 | 1011 | 0 |
| 12 | 1100 | 1 |
| 13 | 1101 | 0 |
| 14 | 1110 | 0 |
| 15 | 1111 | 1 |

A single parity bit can detect the existence of an error if only one error occurs, or if the number of errors is odd. If an even number of errors occurs, they will go undetected by this scheme. Although there is no significant advantage to using odd or even parity, odd parity is often used because it detects the absence of any signal (the all-zero state).

## ERROR CORRECTING CODES

Parity checking, forbidden combinations of BCD codes, and the use of gray code are several of the methods which can detect when an error has occurred. Nevertheless, these methods do not determine what the error is and correct it. There is a class of codes called error correcting codes which does perform this function.

### Hamming Code

The **Hamming** code is the most widely used single-error correcting code. Basically, the Hamming code is a multiple parity scheme. For a given number of bits, there is an associated number of even parity bits. These parity bits check different groupings of the data bits for even parity, each parity bit checking a separate grouping. An example of the Hamming code applied to a 4-bit binary number is shown in Table 2-5. From the table, it can be verified that the three parity bits—P1, P2, and P3—are associated with the data bits as follows: P1-8-4-1, P2-8-2-1, P3-4-2-1. Each parity bit represents even parity for its associated three data bits. Comparison of these three parity bits will not only indicate when there is an error, but will also indicate which bit is wrong.

Table 2-5. Hamming Code for Four Data Bits.

| DECIMAL NUMBER | 8 | 4 | 2 | 1 | P1 | P2 | P3 |
|---|---|---|---|---|---|---|---|
| 0 | 0 | 0 | 0 | 0 | 0 | 0 | 0 |
| 1 | 0 | 0 | 0 | 1 | 1 | 1 | 1 |
| 2 | 0 | 0 | 1 | 0 | 0 | 1 | 1 |
| 3 | 0 | 0 | 1 | 1 | 1 | 0 | 0 |
| 4 | 0 | 1 | 0 | 0 | 1 | 0 | 1 |
| 5 | 0 | 1 | 0 | 1 | 0 | 1 | 0 |
| 6 | 0 | 1 | 1 | 0 | 1 | 1 | 0 |
| 7 | 0 | 1 | 1 | 1 | 0 | 0 | 1 |
| 8 | 1 | 0 | 0 | 0 | 1 | 1 | 0 |
| 9 | 1 | 0 | 0 | 1 | 0 | 0 | 1 |
| 10 | 1 | 0 | 1 | 0 | 1 | 0 | 1 |
| 11 | 1 | 0 | 1 | 1 | 0 | 1 | 0 |
| 12 | 1 | 1 | 0 | 0 | 0 | 1 | 1 |
| 13 | 1 | 1 | 0 | 1 | 1 | 0 | 0 |
| 14 | 1 | 1 | 1 | 0 | 0 | 0 | 0 |
| 15 | 1 | 1 | 1 | 1 | 1 | 1 | 1 |

The mechanism for determining and correcting errors lies in the regeneration of parity at the receiving end. Three new parity bits are generated for three 4-bit groups which now include the

Table 2-6. Hamming Code Regrouped for Error Correction.

| DECIMAL NUMBER | P1 | P2 | 8 | P3 | 4 | 2 | 1 |
|---|---|---|---|---|---|---|---|
| 0 | 0 | 0 | 0 | 0 | 0 | 0 | 0 |
| 1 | 1 | 1 | 0 | 1 | 0 | 0 | 1 |
| 2 | 0 | 1 | 0 | 1 | 0 | 1 | 0 |
| 3 | 1 | 0 | 0 | 0 | 0 | 1 | 1 |
| 4 | 1 | 0 | 0 | 1 | 1 | 0 | 0 |
| 5 | 0 | 1 | 0 | 0 | 1 | 0 | 1 |
| 6 | 1 | 1 | 0 | 0 | 1 | 1 | 0 |
| 7 | 0 | 0 | 0 | 1 | 1 | 1 | 1 |
| 8 | 1 | 1 | 1 | 0 | 0 | 0 | 0 |
| 9 | 0 | 0 | 1 | 1 | 0 | 0 | 1 |
| 10 | 1 | 0 | 1 | 1 | 0 | 1 | 0 |
| 11 | 0 | 1 | 1 | 0 | 0 | 1 | 1 |
| 12 | 0 | 1 | 1 | 1 | 1 | 0 | 0 |
| 13 | 1 | 0 | 1 | 0 | 1 | 0 | 1 |
| 14 | 0 | 0 | 1 | 0 | 1 | 1 | 0 |
| 15 | 1 | 1 | 1 | 1 | 1 | 1 | 1 |

transmitted parity bits. By regrouping the seven received bits as shown in Table 2-6, the result indicated by the formation of the new parity bits locates the bit position in which an error occurred, if any. For example, consider an error in transmitting the number 13 (1010101) such that the code is received (1010111). The three new parity bits are computed as follows:

| Received | P1 | P2 | 8 | P3 | 4 | 2 | 1 |
|---|---|---|---|---|---|---|---|
| Code | 1 | 0 | 1 | 0 | 1 | 1 | 1 |

New P1 for (P1-8-4-1) = 0
New P2 for (P2-8-2-1) = 1
New P3 for (P3-4-2-1) = 1

The binary number $P3P2P1 = 110$ indicates that an error occurred in bit 6 counting from the left. A correction (inversion) can now be performed on this bit. If the check number $P3P2P1 = 000$, then the given number is correct. Obviously, only one error can be corrected with this code and it is thus labeled a single-error correcting code. If more check bits were added, it would be possible to correct more than one error.

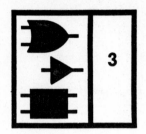

3

# Binary Arithmetic

The various binary number representations and binary codes all have unique characteristics which make them convenient in some particular application. For pure binary bits, there are very simple rules which always apply when adding, subtracting, multiplying, or dividing. But when bits are combined into one of the specialized notations described in Chapters 1 and 2, care must be taken to recognize the proper notation and to treat these bits accordingly. For example, 10111 (which stands for 23 in pure binary) means −7 in sign magnitude notation, −8 in one's complement notation, −9 in two's complement notation, 17 in 8421 BCD, and 14 in excess-three BCD. Clearly, binary arithmetic operations must be varied to suit each particular number representation format or scheme being used.

## BASIC RULES FOR BINARY ARITHMETIC

The binary number system has the same basic format as any other number system, and thus, arithmetic operations are performed in the same manner as in other systems. For example, in the binary system, 0+0=0, 0+1=1, and 1+0=1. However, since there is no single binary digit for the number 2, 1+1=0 and there is a 1 to carry. This carry is similar to a decimal carry where the decimal sum exceeds the number 9. Table 3-1 shows addition, subtraction, multiplication, and division examples for the binary number system.

As can be seen from the table, multiplication and division are generally the most complex arithmetic operations and require several steps to perform. But if the number to be used as a multiplier or divisor happens to be a power of two, the operation becomes extremely simple. To multiply by a power of two, merely shift the number being multiplied by a number of digits equal to the power of two being multiplied by. Consider the multiplication $13 \times 8 = 104$. Then, the multiplication is;

$$13 \quad \times \quad 8 \quad = \quad 104$$
$$\lfloor 1\ 1\ 0\ 1 \rfloor \times 1\ 0\ 0\ 0 = \lfloor 1\ 1\ 0\ 1 \rfloor\ 0\ 0\ 0 \rfloor$$

Original Number ——————— Three place left shift

Similarly, if a number happens to be divided into a power of two, a right shift by the proper number of digits will perform the division. Suppose, for example, that the decimal number thirteen is to be divided by four, $13/4 = 3.25$, again noting that 4 is the second power of 2. The division is:

$$13 \quad \div \quad 4 \quad = \quad 3.25$$
$$\lfloor 1\ 1\ 0\ 1 \rfloor \div 1\ 0\ 0 = \lfloor 0\ 0 \lfloor 1\ 1\ .\ 0\ 1 \rfloor$$

Original Number ———— Two place right shift

## ARITHMETIC WITH SIGN MAGNITUDE NOTATION

Addition and subtraction of sign magnitude number follows the basic rules given for the binary number system but included with each number is a sign bit. The sign bit is actually another plus or minus sign, where plus is denoted by 0 and minus is denoted by 1. As in decimal arithmetic, extreme care must be exercised to make sure all signs are accounted for. For example, $(+7)+(-7)=0$ in the decimal number system. Similarly, $0\ 0111+1\ 0111=0\ 0000$ is sign magnitude notation. An example of a sign magnitude addition is shown below. Note that due to the sign bits, a subtraction is actually performed just as a subtraction is performed for the decimal addition shown at the side. Also, note that two separate operations are required for sign magnitude notation, addition and subtractions. Both *one's complement* notation and *two's complement* notation require only one operation: addition.

Sign Magnitude Addition Example

$$
\begin{array}{rr}
+\ 14 & 0\ \ 1110 \\
+\ (-7) & +\ \underline{1\ \ 0111} \\
\hline
+\ 7 & 0\ \ 0111
\end{array}
$$

Multiplication of sign magnitude numbers is done exactly as in Table 3-1, using the sign bits only for determining the sign of the product. Since multiplication is a process involving a number

**Table 3-1. Binary Arithmetic.**

| BINARY ADDITION |
|---|

| | | |
|---|---|---|
| $0 + 0 = 0$ | | |
| $0 + 1 = 1$ | 13 | 01101 |
| $1 + 0 = 1$ | $+\,14$ | 01110 |
| $1 + 1 = 0$, carry 1 | 27 | 11011 |

| BINARY SUBTRACTION |
|---|

| | | |
|---|---|---|
| $0 - 0 = 0$ | | |
| $0 - 1 = 1$, borrow 1 | 27 | 11011 |
| $1 - 0 = 1$ | $-\,11$ | 01011 |
| $1 - 1 = 0$ | 16 | 10000 |

| BINARY MULTIPLICATION |
|---|

| | | |
|---|---|---|
| $0 \times 0 = 0$ | | 10101 |
| $0 \times 1 = 0$ | | 101 |
| $1 \times 0 = 0$ | 21 | 10101 |
| $1 \times 1 = 1$ | $\times\ 5$ | 00000 |
| | 105 | 10101 |
| | | 1101001 |

| BINARY DIVISION |
|---|

| | | |
|---|---|---|
| $0 \div 0 = ?$ | 21 | 10101 |
| $0 \div 1 = 0$ | 5)105 | 101)1101001 |
| $1 \div 0 = ?$ | 10 | 101 |
| $1 \div 1 = 1$ | 5 | 110 |
| | 5 | 101 |
| | | 101 |
| | | 101 |

of steps, an actual procedure used by digital machines will be examined here to show how the individul operations are performed. Sign magnitude multiplication can be accomplished by repeatedly adding either the number being multiplied or zero, and shifting one place to the right after each operation. Consider the example shown in Table 3-2. The multiplier bits have been labeled B2, B1, and B0 for convenience. Bit B2 is the most significant bit of the multiplier and bit B0 is the least significant bit. Each time the multiplier bit is 1, 10101 is added and a right shift is performed. When the multiplier bit is 0, 00000 is added and a right shift is performed. At each step, an accumulated result is maintained and the process is continued until all bits of the multiplier have been used as operators.

**Table 3-2. Example of Sign Magnitude Multiplication.**

Problem: 10101 x 101 = ?

| Multiplier B2B1B0 = 101 | OPERATION | ACCUMULATIVE RESULT |
|---|---|---|
| B0 = 1 (LSB) | Add 10101 | 00000    (Starting point = 0)<br>+ 10101<br>10101 |
| B1 = 0 | Shift right<br>Do nothing<br>(Add 00000) | 010101<br>− 00000<br>010101 |
| B2 = 1 (MSB) | Shift right<br>Add 10101 | 0010101<br>└ 10101<br>1101001  (Final product) |

Result: 10101 x 101 = 1101001
21 x 5 = 105
NOTES: A right shift is not required after the last addition.

**Operation Codes**
Multiplier bit = 0  Do nothing (Add 0)
Multiplier bit = 1  Add

At the final step, no further right shift is required. There is nothing special about the two numbers used in this example; indeed, the procedure described works with any two **sign magnitude** numbers, either number consisting of as many bits as desired. Since the sign bits are not used for the actual multiplication procedure, it is apparent that the method works equally well for a pure binary number.

A sign magnitude division uses repeated subtractions and left shifts to obtain its results. The division process is usually somewhat more complicated than multiplication, in that a division will not necessarily produce a specific number of digits. Just as one divided by three, in decimal, results in an endless string of three's ($\frac{1}{3} = 0.333333333...$), there are many binary fractions which can be formed where the division, when performed, also produces a very large number of digits. Therefore, a trial-and-error procedure is required, in which repeated division operations are performed until the desired accuracy is obtained or until a remainder of zero occurs. The division is terminated when the desired number of digits is obtained. Of course, any time a remainder of zero occurs, the division is complete and the resultant answer is exact.

Table 3-3 shows an example of a typical sign magnitude division. Initially, note that the divisor is smaller than the

### Table 3-3. Example of Sign Magnitude Division.

Problem: 1101001 ÷ 101 = ?

| OPERATION | ACCUMULATIVE RESULT | RESULT ≥ DIVISOR ? | QUOTIENT |
|---|---|---|---|
| | 01101001 (starting point) | — | |
| Shift left<br>Subtract | 11010010<br>− 10100000<br>00110010 | Yes | 1 (MSB) |
| Shift left<br>Do nothing | 01100100<br>− 00000000<br>01100100 | No | 0 |
| Shift left<br>Subtract | 11001000<br>− 10100000<br>00101000 | Yes | 1 |
| Shift left<br>Do nothing | 01010000<br>− 00000000<br>01010000 | No | 0 |
| Shift left<br>Subtract | 10100000<br>− 10100000<br>00000000 | Yes | 1 (LSB) |

Tentative answer 0.10101
Final answer 10101
Result: 1101001 ÷ 101 = 10101

NOTES: At start of problem divisor must be greater than number being divided. If not, shift left until it is. At end of problem, shift answer same number of places.

**OPERATION CODES**

| RESULT ≥ DIVISOR ? | OPERATION | QUOTIENT |
|---|---|---|
| Yes<br>No | Subtract<br>Do nothing<br>(Subtract 0) | 1<br>0 |

number being divided. For this type of binary division, it is required that the divisor be the larger number. Hence, a five-place left shift is performed to increase the size of the divisor, retaining this information for later use.

The division process is begun by writing down the number to be divided, in this case 1101001. For each step of the process, a left shift is performed then the new number is examined to determine whether or not it is larger than the divisor. If it is, the quotient is a 1 for that bit and the divisor is subtracted from the accumulative result.

If the new number is smaller than the divisor, the quotient is a 0 for that bit and 00000000 is subtracted from the accumulative result.

This shift and subtract procedure is repeated again and again until finally enough digits have been generated for the desired quotient accuracy or until an accumulative result of zero occurs. The final quotient is written down in the form 0.xxxxxxxx, with a left shift to be performed equal to the number of places of left shift used when starting the problem. In this example, the quotient is initially written 0.10101 then shifted left five places, making the final answer 10101.

## ARITHMETIC WITH ONE'S COMPLEMENT NOTATION

One's complement addition is just like straight binary addition, except that there is an added bit, the sign bit. The sign bit is included as one of the bits which must be added. When all bits are added in the normal binary fashion, a positive or negative one's complement number is obtained by including the carry bit as part of the addition. If the carry bit is a zero, the answer is correct and no further correction is required. If the carry bit is a one, then a one must be added to the sum to obtain a correct answer. The process of adding the carry bit to the sum is called **end-around carry**. An example of one's complement addition is shown below.

$$
\begin{array}{r}
+\ 14 \\
+\ (-7) \\
\hline
+\ 7 \quad \boxed{1}
\end{array}
\qquad
\begin{array}{r}
0\ 1\ 1\ 1\ 0 \\
+\ 1\ 0\ 0\ 0\ 0 \\
\hline
0\ 0\ 1\ 1\ 0 \\
\longrightarrow 1 \\
\hline
0\ 0\ 1\ 1\ 1
\end{array}
$$

One of the main advantages to one's complement numbers is that subtraction is never required. If a subtraction is called for, the number being subtracted is simply inverted (complemented) then the two numbers are added. Exactly the same rules apply as for the additon just described. Addition in place of subtraction is a great convenience, and if the end-around carry could also be eliminated, the arithmetic would be further simplified. Two's complement numbers, considered next, provide this feature. An example of subtraction of one's complement numbers is shown at the top of the next page.

$$
\begin{array}{r} +\ 10 \\ -\ (-5) \\ \hline +\ 15 \end{array}
\qquad
\begin{array}{r} 0\ \ 1\ 0\ 1\ 0 \\ -\ \ 1\ \ 1\ 0\ 1\ 0 \\ \hline \end{array}
\ \rightarrow\
\begin{array}{r} 0\ \ 1\ 0\ 1\ 0 \\ +\ \ 0\ \ 0\ 1\ 0\ 1 \\ \hline 0\ \ 1\ 1\ 1\ 1 \end{array}
$$

## ARITHMETIC WITH TWO'S COMPLEMENT NOATION

Two's complement is the most widely used notation where arithmetic operations are required. It is extremely easy to use for addition and subtraction and it lends itself readily to the multiplication and division operations. The addition of two's complement numbers is just like one's complement addition, except that the end-around carry is not required. All bits including the sign bits are added and the result is the sum. Any carry is ignored. An example of two's complement addition is shown below.

$$
\begin{array}{r} +\ 14 \\ +\ (-7) \\ \hline +\ 7 \end{array}
\qquad
\begin{array}{r} 0\ \ 1\ 1\ 1\ 0 \\ +\ \ 1\ \ 1\ 0\ 0\ 1 \\ \hline 0\ \ 0\ 1\ 1\ 1 \end{array}
$$

Two's complement numbers, like one's complement numbers, are subtracted by complementing,then adding.It should be remembered that two's complements are formed by inverting and adding 1. Thus, to subtract any two numbers, invert the number being subtracted, then add the two numbers, then add 1. When adding the two numbers, the same rules apply as for the two's complement addition just described. An example of two's complement subtraction is shown below.

$$
\begin{array}{r} +\ 10 \\ -\ (-5) \\ \hline +\ 15 \end{array}
\qquad
\begin{array}{r} 0\ \ 1\ 0\ 1\ 0 \\ -\ \ 1\ \ 1\ 0\ 1\ 1 \\ \hline \end{array}
\ \rightarrow\
\begin{array}{r} 0\ \ 1\ 0\ 1\ 0 \\ +\ \ 0\ \ 0\ 1\ 0\ 0 \\ \hline 0\ \ 1\ 1\ 1\ 0 \\ 1 \\ \hline 0\ \ 1\ 1\ 1\ 1 \end{array}
$$

The multiplication of two numbers in two's complement notation is a modification of the shift-and-add technique utilized for sign magnitude multiplication. Here, the multiplier forms an operation code which tells whether to add, subtract, or do nothing for each step of the procedure. After each operation, a right shift occurs, then the next operation is performed.

Table 3-4 shows a typical two's complement multiplication. Two bit groupings of the multiplier form the operation codes

**Table 3-4. Example of Two's Complement Multiplication.**

Problem: 0 10101 x 0 101 = ?

| MULTIPLIER | OPERATION CODE | OPERATION | ACCUMULATIVE RESULT |
|---|---|---|---|
| 0 1 0[1 -] | 10 | Subtract 0 10101 | 0 00000   (Strating point = 0)<br>+ 1 01011<br>‾‾‾‾‾‾‾‾<br>1 01011 |
| 0 1[0 1] | 01 | Shift right<br>Add 0 10101 | 1 101011<br>+ 0 10101<br>‾‾‾‾‾‾‾‾<br>0 010101 |
| 0 [1 0]1 | 10 | Shift right<br>Subtract 0 10101 | 0 0010101<br>+ 1 01011<br>‾‾‾‾‾‾‾‾<br>1 1000001 |
| [0 1]0 1 | 01 | Shift right<br>Add 0 10101 | 1 11000001<br>+ 0 10101<br>‾‾‾‾‾‾‾‾<br>0 01101001  (Final product) |

NOTES: A right shift is not required after the last operation.

Result: 0 10101 x 0 101 = 0 1101001<br>+21 x +5 = +105

When shifting two's complement numbers to the right, the new bit shifted in is the same as the sign bit.

**Operation Codes**
00   Do nothing
01   Add
10   Subtract
11   Do nothing

which are interpreted as shown in the notes. After each operation, a right shift is accomplished, being careful to shift in the correct polarity bit as determined by the sign of the number. All addition and subtractions follow the previously established rules for two's complement arithmetic.

Comparison of this technique to the one shown in Table 3-2 reveals that the two methods are quite similar, except that two bit groupings are used to form the operation codes, and that subtraction is one of the operations allowed for the two's complement case.

Division of two's complement numbers is also quite similar to the sign magnitude technique. The main difference is that the sign of the divisor and the sign of the result serve together to determine the operation code for each step. Also, addition and subtraction are both permitted operations for two's complement division, whereas the operations for sign magnitude division were limited to either subtract or do nothing (subtract 0). An example of two's complement division is shown in Table 3-5. As with sign magnitude division, the starting point for the process is with the number being divided. This is followed by a left shift, then the

## Table 3-5. Example of Two's Complement Division.

Problem: 0 1101001 ÷ 0 101 = ?

| OPERATION | ACCUMULATIVE RESULT | SIGN SAME AS DIVISOR? | QUOTIENT |
|---|---|---|---|
| | 0 01101001 (Starting point) | − | |
| Shift left<br>Subtract | 0 11010010<br>+ 1 01100000<br>0 00110010 | Yes | 1 (MSB) |
| Shift left<br>Subtract | 0 01100100<br>+ 1 01100000<br>1 11000100 | Yes | 1 |
| Shift left<br>Add | 1 10001000<br>+ 0 10100000<br>0 00101000 | No | 0 |
| Shift left<br>Subtract | 0 01010000<br>+ 1 01100000<br>1 10110000 | Yes | 1 |
| Shift left<br>Add | 1 01100000<br>+ 0 10100000<br>0 00000000 | No | 0 (LSB) |

| | | |
|---|---|---|
| Tentative Answer | 1 | .1010 |
| Correction factor | 1 | 00001 |
| | 0 | .10101 |
| Shifted | 0 | 10101 |

Result: 0 1101001 ÷ 0 101 = 0 10101

NOTES. At start of problem, divisor must be greater than number being divided. If not, shift left until it is. At end of problem, shift answer same number of places.

Correction factor is always 1 000...001 with last one being one place more than tentative answer.

**OPERATION CODES**

| SIGN SAME AS DIVISOR? | OPERATION | QUOTIENT |
|---|---|---|
| Yes<br>No | Subtract<br>Add | 1<br>0 |

indicated operation. The procedure is repeated until a result of 0 is obtained or until the desired number of digits have been generated.

Of importance with the two's complement division technique is the fact that the answer obtained is initially incorrect. To get the correct answer, 1    000...001 must be added to the quotient. This correction can be thought of as simply changing the sign of the answer, then adding a one at the end. Comparison of Tables 3-3 and 3-5 shows the similarities in the two techniques.

# ARITHMETIC OPERATIONS WITH BCD CODES

A number represented in a BCD code is organized into four bit groups called decades, with each decade representing one digit of a decimal number. Within each decade, the rules for addition and subtraction are identical to the basic rules shown in Table 3-1. In standard binary notation, each adjacent binary bit represents an increasing power of two; carries or borrows are normal binary functions as previously described. In BCD, however, each decade uses only 10 of the possible 16 states available. As a result, special correction factors must be added or subtracted when using BCD to account for the unused states. Further, carries and borrows external to each decade must be based on the requirements for a decimal carry or borrow instead of a binary carry or borrow.

## Addition with the 8421 Code

A common method of performing BCD addition is to add two numbers in binary fashion, and then add or subtract an appropriate correction factor, if necessary. This is the method described here.

The initial addition will be correct using the 8421 code, provided the sum is not greater than 9. To take care of cases where the sum does exceed 9, 6 must be added to convert the number to a corrected code. Thus, if the decimal sum is between 10 and 15, 6 must be added to the initial result and a carry generated for the next decade. If the decimal sum is greater than 15, a carry is automatically generated; however, it is still necessary to add 6 to the initial result. Table 3-6 indicates the

Table 3-6. Corrections for Addition Using 8421 Code.

| DECIMAL SUM | UNCORRECTED | | CORRECTED | | CORRECTION |
|---|---|---|---|---|---|
| | CARRY | SUM | CARRY | SUM | |
| 10 | 0 | 1010 | 1 | 0000 | +6 |
| 11 | 0 | 1011 | 1 | 0001 | +6 |
| 12 | 0 | 1100 | 1 | 0010 | +6 |
| 13 | 0 | 1101 | 1 | 0011 | +6 |
| 14 | 0 | 1110 | 1 | 0100 | +6 |
| 15 | 0 | 1111 | 1 | 0110 | +6 |
| 16 | 1 | 0000 | 1 | 0111 | +6 |
| 17 | 1 | 0001 | 1 | 1000 | +6 |
| 18 | 1 | 0010 | 1 | 1001 | +6 |
| 19 | 1 | 0011 | 1 | 0101 | +6 |

corrections which must be made when performing addition with the 8421 code.

One disadvantage to correcting the 8421 code in the manner described is that each decade must be corrected, starting with the least significant decade, before the next decade can be corrected. The example below demonstrates addition using the 8421 code.

Single Decade Addition

```
    9          1 0 0 1
    7          0 1 1 1
⎣1⎦ 6    1     0 0 0 0
 ↑             1 1 0
 •      →⎣1⎦    0 1 1 0
```

Multiple Decimal Decade Addition Carry

```
349    0 0 1 1    0 1 0 0    1 0 0 1
163    0 0 0 1    0 1 1 0    0 0 1 1
512    0 1 0 0    1 0 1 0    1 1 0 0
                  1 1 0      1 1 0
                        ⎣1⎦←
              ⎣1⎦←
       0 1 0 1    0 0 0 1    0 0 1 0
        (5)        (1)        (2)
```

**Addition with the Excess-Three Code.**

When two excess-three numbers are added, the sum of the two excess 3's results in an excess 6. If the resultant decimal sum is 9 or less, a correction of − 3 must be made to return to proper excess-three notation. However, if the decimal sum is greater than 9, the excess 6 will be larger than 15, thereby generating a carry from the decade and causing an **overflow** within the decade. The term "overflow" refers to the situation where the sum requires more bits than are available within the decade. When this occurs, the sum shown by the 4 bits in the decade is incorrect and the carry bit must be sensed to properly correct the sum. Thus, in this case a correction of + 3 is required to return excess-three notation. Table 3-7 shows the corrections to be made when performing addition with the excess-three code.

A convenient feature of the excess-three code is that the carry generated from the initial addition can be used to determine whether a + 3 or a − 3 correction is to be made. If there is no carry, then − 3 must be added. If a carry does occur, + 3 is to be added. The example below demonstrates addition using the excess-three code.

Single Decade Addition

```
    9          1 1 0 0
    7          1 0 1 0
⎣1⎦ 6    1     0 1 1 0
 ↑           + 0 0 1 1
 •      →⎣1⎦    1 0 0 1
```

Multiple Decade Addition

```
349    0 1 1 0    0 1 1 1    1 1 0 0
163    0 1 0 0    1 0 0 1    0 1 1 0
512    1 0 1 0    0 0 0 0    0 0 1 0
     − 0 0 1 1  + 0 0 1 1  + 0 0 1 1
         1          1
       1 0 0 0    0 1 0 0    0 1 0 1
        (5)        (1)        (2)
```

Decimal Carry

42

## Subtraction with the 8421 Code

The results for a subtraction using the 8421 code will be correct, providing the difference obtained is a positive number. However, if the difference is negative a borrow must be performed from the next decade and a correction of $-6$ must be made to the binary difference to obtain the correct result. Table 3-8 lists the corrections to be made for subtraction using the 8421 code.

As with 8421 addition, each decade must be corrected in order, starting with the least significant decade. An example of subtraction using 8421 code is shown below.

$$
\begin{array}{r}
45 \\
-\ 28 \\
\hline
17
\end{array}
\qquad
\begin{array}{r}
0\ 1\ 0\ 0 \\
-\ 0\ 0\ 1\ 0 \\
\hline
0\ 0\ 1\ 0 \\
-\ 1 \\
\hline
0\ 0\ 0\ 1 \\
(1)
\end{array}
\qquad
\begin{array}{r}
0\ 1\ 0\ 1 \\
1\ 0\ 0\ 0 \\
\hline
1\ 1\ 0\ 1 \\
-\ 0\ 1\ 1\ 0 \\
\hline
0\ 1\ 1\ 1 \\
(7)
\end{array}
$$

## Subtraction with the Excess-Three Code.

An excess-three subtraction is similar to an 8421 subtraction, except that different correction factors are required. If the difference is positive, the correction required is $+3$ and if the difference is negative, then a correction of $-3$ is needed. Table

Table 3-7. Corrections for Addition using Excess Three Code.

| DECIMAL SUM | UNCORRECTED | | CORRECTED | | CORRECTION |
|---|---|---|---|---|---|
| | CARRY | SUM | CARRY | SUM | |
| 0 | 0 | 0110 | 0 | 0011 | −3 |
| 1 | 0 | 0111 | 0 | 0100 | −3 |
| 2 | 0 | 1000 | 0 | 0101 | −3 |
| 3 | 0 | 1001 | 0 | 0110 | −3 |
| 4 | 0 | 1010 | 0 | 0111 | −3 |
| 5 | 0 | 1011 | 0 | 1000 | −3 |
| 6 | 0 | 1100 | 0 | 1001 | −3 |
| 7 | 0 | 1101 | 0 | 1010 | −3 |
| 8 | 0 | 1110 | 0 | 1011 | −3 |
| 9 | 0 | 1111 | 0 | 1100 | −3 |
| 10 | 1 | 0000 | 1 | 0011 | +3 |
| 11 | 1 | 0001 | 1 | 0100 | +3 |
| 12 | 1 | 0010 | 1 | 0101 | +3 |
| 13 | 1 | 0011 | 1 | 0110 | +3 |
| 14 | 1 | 0100 | 1 | 0111 | +3 |
| 15 | 1 | 0101 | 1 | 1000 | +3 |
| 16 | 1 | 0110 | 1 | 1001 | +3 |
| 17 | 1 | 0111 | 1 | 1010 | +3 |
| 18 | 1 | 1000 | 1 | 1011 | +3 |
| 19 | 1 | 1001 | 1 | 1100 | +3 |

**Table 3-8. Corrections for Subtraction Using 8421 Code.**

| DECIMAL DIFFERENCE | UNCORRECTED | | CORRECTED | | CORRECTION |
|---|---|---|---|---|---|
| | BORROW | DIFFERENCE | BORROW | DIFFERENCE | |
| −1 | 1 | 1111 | 1 | 1001 | −6 |
| −2 | 1 | 1110 | 1 | 1000 | −6 |
| −3 | 1 | 1101 | 1 | 0111 | −6 |
| −4 | 1 | 1100 | 1 | 0110 | −6 |
| −5 | 1 | 1011 | 1 | 0101 | −6 |
| −6 | 1 | 1010 | 1 | 0100 | −6 |
| −7 | 1 | 1001 | 1 | 0011 | −6 |
| −8 | 1 | 1000 | 1 | 0010 | −6 |
| −9 | 1 | 0111 | 1 | 0001 | −6 |

3-9 shows the corrections to be made for subtraction using the excess-three code.

From Table 3-9, it can be seen that a borrow is generated if the difference is negative and that there is no borrow when the difference is positive. An example of subtraction using excess-three code is shown below.

$$
\begin{array}{r}
45 \\
-\ 28 \\
\hline
17
\end{array}
\qquad
\begin{array}{r}
0\ 1\ 1\ 1 \\
-\ 0\ 1\ 0\ 1 \\
\hline
0\ 0\ 1\ 0 \\
-\ 1 \\
+\ 0\ 0\ 1\ 1 \\
\hline
0\ 1\ 0\ 0 \\
(1)
\end{array}
\qquad
\begin{array}{r}
1\ 0\ 0\ 0 \\
1\ 0\ 1\ 1 \\
\hline
1\ 1\ 0\ 1 \\
\\
-\ 0\ 0\ 1\ 1 \\
\hline
1\ 0\ 1\ 0 \\
(7)
\end{array}
$$

**Table 3-9. Corrections for Subtraction Using Excess Three Code.**

| DECIMAL DIFFERENCE | UNCORRECTED | | CORRECTED | | CORRECTION |
|---|---|---|---|---|---|
| | BORROW | DIFFERENCE | BORROW | DIFFERENCE | |
| +9 | 0 | 1001 | 0 | 1100 | +3 |
| +8 | 0 | 1000 | 0 | 1011 | +3 |
| +7 | 0 | 0111 | 0 | 1010 | +3 |
| +6 | 0 | 0110 | 0 | 1001 | +3 |
| +5 | 0 | 0101 | 0 | 1000 | +3 |
| +4 | 0 | 0100 | 0 | 0111 | +3 |
| +3 | 0 | 0011 | 0 | 0110 | +3 |
| +2 | 0 | 0010 | 0 | 0101 | +3 |
| +1 | 0 | 0001 | 0 | 0100 | +3 |
| 0 | 0 | 0000 | 0 | 0011 | +3 |
| −1 | 1 | 1111 | 1 | 1100 | −3 |
| −2 | 1 | 1110 | 1 | 1011 | −3 |
| −3 | 1 | 1101 | 1 | 1010 | −3 |
| −4 | 1 | 1100 | 1 | 1001 | −3 |
| −5 | 1 | 1011 | 1 | 1000 | −3 |
| −6 | 1 | 1010 | 1 | 0111 | −3 |
| −7 | 1 | 1001 | 1 | 0110 | −3 |
| −8 | 1 | 1000 | 1 | 0101 | −3 |
| −9 | 1 | 0111 | 1 | 0100 | −3 |

# Logic Fundamentals

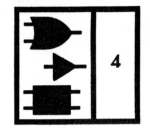

Digital circuits perform the binary arithmetic operations described in Chapters 1, 2, and 3 with the binary digits 1 and 0. Similarly, there are other types of functions which can be performed with digital circuits, using only these two binary numerals. These functions are referred to as **logic** functions. Logic functions can all be described in terms of algebraic statements which are either true or false. The **true** condition is represented by the binary digit 1, and the **false** condition by the binary digit 0.

## BOOLEAN ALGEBRA

The algebra used to symbolically describe logic functions is called Boolean algebra. As with ordinary algebra, the letters of the alphabet can be used to represent variables, the primary difference being that Boolean algebra variables can only have the values of 1 or 0. There are three connecting symbols used in Boolean algebra. These are the **equal** sign ( = ), the **plus** sign ( + ), and the **multiply** symbol (·). They are defined as follows:

(1) The equal sign ( = ) refers to a standard mathematical equality. That is, the logical value on one side of the sign is identical to the logical value on the other side of the sign. Hence, given that there are two logical variables, A and B, such that $A = B$, it follows that if $A = 1$ then $B = 1$, and if $A = 0$ then $B = 0$.

(2) The plus sign ( + ) in Boolean algebra refers to the logical OR function. If $A + B = 1$, the meaning is that either A or B represents the logical value 1. Therefore, when the statement $A + B = 1$ appears, then either $A = 1$ or $B = 1$ or both.

(3) The logical multiply sign (·) is called the AND function. If it is given that $A·B = 1$, the interpretation is that both A and B represent the logical value 1. Thus, an algebraic equation of the form $A·B = 1$ means that $A = 1$ and $B = 1$. If either A or

**45**

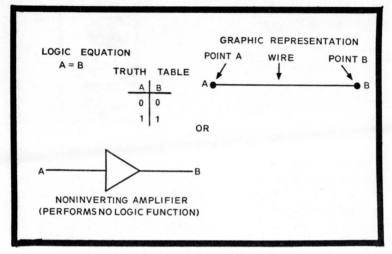

**LOGIC EQUATION**
A = B

**GRAPHIC REPRESENTATION**

POINT A      WIRE      POINT B

A ●————————————————● B

**TRUTH TABLE**

| A | B |
|---|---|
| 0 | 0 |
| 1 | 1 |

OR

A ————————▷———————— B

NONINVERTING AMPLIFIER
(PERFORMS NO LOGIC FUNCTION)

Fig. 4-1. Logical equality.

B is not a logical 1, then AB cannot possibly be a logical 1.
The function A·B is often written AB, omitting the dot for convenience.

Boolean algebra also uses a logical **NOT** operation, indicated by a bar over the variable. The NOT has the effect of inverting (complementing) the logical value. Thus, if $A = 1$, then $\bar{A} = 0$. Additionally, parentheses are frequently used, as in regular algebra, to indicate groupings of logical operations.

Used in conjunction with Boolean algebra are truth tables and logic symbols. A truth table simply shows all of the possible values for the inputs to a function, then shows the resultant output for each combination of inputs. Of course, the purpose of truth tables is not to determine whether or not circuits are "lying." Rather, they verify "truth" in the logical sense—that a given set of input conditions will produce a "true" output, while some other set of input conditions produces a **false** output. As was previously stated, a true state is indicated by the digit 1, and a false state is indicated by the digit 0. A logic symbol is a graphic way of indicating a particular logic function.

## LOGIC FUNCTION

Since Boolean algebra is based on elements having two possible stable states, it becomes very useful for analyzing logic (switching) circuits. The reason for this is that a switching circuit can be in only one of two possible states; that is, it is either a closed circuit or an open circuit—it's either turned on or it's turned off. As stated earlier, these two states can easily be represented by the binary values of 1 and 0.

**Logical Equality**

The logic function called equality actually means that two points or two signals are functionally identical. Referring to Fig. 4-1, it is seen that a piece of wire or some arbitrary buffer amplifier is a typical condition of equality in a digital circuit. The logic condition at point A is identical to the logic condition at point B. Also, the truth table shows the relationship of input to output conditions. Namely, if A = 0 then B = 0, and if A = 1 then B = 1. Recalling, this is exactly the definition for equality given in the preceding paragraphs defining the basic functions for Boolean algebra

**OR Function**

To clarify the meaning of an OR function, consider the two switches A and B, shown in Fig. 4-2. Both switches are connected to some positive voltage, + V. it is seen that the positive voltage will be present at point F if switch A is closed or if switch B is closed. If neither switch is closed, the positive voltage will not be present at point F. The truth table tells the same story in terms of ones and zeroes. For the A and B columns, one means a closed switch and zero means an open switch. In the F column, one refers to the positive voltage being present and zero refers to the absence of the positive voltage. A comparison of the truth table to the statements given in the OR function definition, shows that the two are just different ways of indicating the same thing.

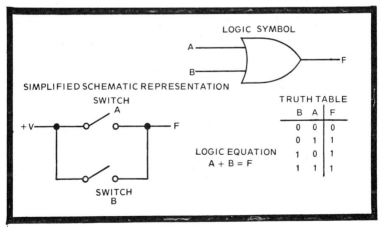

Fig. 4-2. OR function.

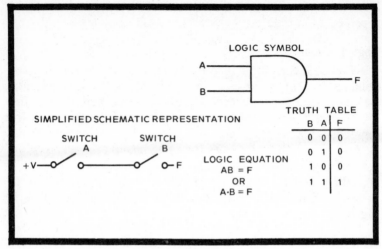

Fig. 4-3. AND function.

## AND Function

The two switches used to represent the OR function were wired in parallel. If, instead, two switches are wired in series as shown in Fig. 4-3, a simplified representation of an AND function is obtained. It is apparent that both switches A and B must be closed to obtain a positive voltage at point F. If either switch is open, the positive voltage will not be present at the output. As was previously the case, an inspection of the truth table shows that the truth table simply reflects the definitions which were previously stated in words.

## NOT Function

The logical NOT operation can be thought of, from a circuit standpoint, as an inverter. Whatever the input to the inverter, the output assumes the opposite polarity. This action is similarly reflected in the truth table. There is no special symbol which separately indicates a NOT operation. Instead, a small circle is attached at the output of the logic symbol for whatever logical function is being performed, and the circle indicates that an inversion of logic levels has taken place. The following paragraphs will further clarify this point.

## NAND and NOR Functions

The AND, OR, and NOT functions were each described with a logic equation, an associated truth table, and a logic symbol. Other functions that are commonly used can be described in a similar fashion. If the AND function of Fig. 4-3 is followed by the inverter of Fig. 4-4, the resultant function is called a NAND (short for NOT-AND) function. From Fig. 4-5, it is seen that the truth table for a NAND function has outputs which are inverted from the outputs shown previously for the AND function. Also, it should be noted that the small circle at the output end of the AND logic symbol (indicating the inverter portion of the circuit) creates a new symbol for a NAND gate. The small circle is frequently used to indicate inversion in logic symbology and wherever it is seen at the output to an AND function, the overall circuit is called a NAND gate. Similarly, a NOR function can be formed by adding an inverter to an OR cirucit. Again the truth table shows that the outputs are inverted from the standard OR circuit and the logic symbol is formed by adding a small circle to the normal OR symbol.

## EXCLUSIVE-OR FUNCTION

The **exclusive-**OR function is another function of interest because it is commonly used in a number of applications. The exclusive-OR merely means the OR of either A or of B but exclusive of each other, thus not including the case of both A and B being true at the same time. The exclusive-OR is represented in equations by the symbol $\oplus$ and is arithmetically equal to modulo-two addition. As will be recalled from Chapter 2, a modulo-two sum is simply a binary sum with no carry. From the above, it can be seen that the exclusive-OR circuit is quite useful where arithmetic operations are to be performed. Also, an exclusive-OR function represents a simple parity circuit. That is, the output of the exclusive-OR is a one if an odd number of bits are one and is a zero if an even number of bits are one. This is identical to the even parity scheme previously described. Odd parity is the inverse of the exclusive-OR function.

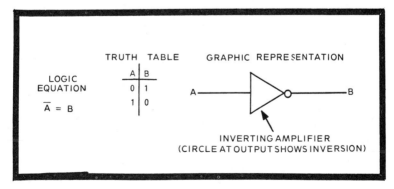

Fig. 4-4. NOT function.

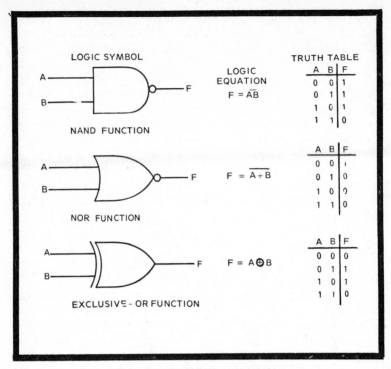

Fig. 4-5. Additional logic functions.

## POSITIVE AND NEGATIVE LOGIC

At the beginning of this chapter, the binary digit 1 was defined to indicate the true logic state and the binary digit zero to indicate the false state. Also, in the circuit examples, the logic 1 was assumed to refer to a positive voltage and the logic zero to a negative voltage. In practice, however, the assignment of the logic 1 and 0 to high or low voltage levels is a matter of personal choice and specific circuit design. For example, a system that uses + 5V could just as well be assigned to a logic 0 and 0V a logic 1.

Since there is this potential confusion over the logic 1 and logic 0 states, it is necessary for one to know in advance which system is used for the circuits being discussed. For reference purposes, the logic system is referred to as **positive-true** logic if the higher voltage represents a logic 1; and it is called **negative-true** logic if the lower voltage represents the logic 1.

Consider again the AND function shown in Fig. 4-3. Assuming positive-true logic, several versions of its truth table are reproduced here for reference.

| A | B | F |
|---|---|---|
| 0V | 0V | 0V |
| 0V | +V | 0V |
| + V | 0V | 0V |
| + V | + V | + V |

Voltage Table

| A | B | F |
|---|---|---|
| L | L | L |
| L | H | L |
| H | L | L |
| H | H | H |

High-Low Table

| A | B | F |
|---|---|---|
| 0 | 0 | 0 |
| 0 | 1 | 0 |
| 1 | 0 | 0 |
| 1 | 1 | 1 |

Logic One-Zero Table

But, supposing that instead of assigning positive-true logic states, the same circuit is designated utilizing a negative-true logic convention. Then, the logic one refers to the low states and the logic zero refers to the high states, as shown below.

| A | B | F |
|---|---|---|
| 0V | 0V | 0V |
| 0V | +V | 0V |
| +V | 0V | 0V |
| + V | +V | +V |

Voltage Table

| A | B | F |
|---|---|---|
| L | L | L |
| L | H | L |
| H | L | L |
| H | H | H |

High-Low Table

| A | B | F |
|---|---|---|
| 1 | 1 | 1 |
| 1 | 0 | 1 |
| 0 | 1 | 1 |
| 0 | 0 | 0 |

Logic One-Zero Table

The voltage table and high-low table in both cases are identical, yet it is apparent from the logic one-zero truth table that an OR function is now being performed. The duality of functions demonstrated here illustrates that any given logic circuit may be considered in either of two logic systems and that the interpretation of the circuit operation is dependent upon the choice of system.

As a general statement, it will be observed that any AND function in one logic system becomes an OR function in the other system. Similarly, and OR function in one system becomes an AND function in the other system. Of course, once the logic system is defined, a given circuit performs only one function within that system (i.e. either AND or OR). Throughout the remainder of this book, positive-true logic will be used, since it is in general easier to understand and avoids the confusion of having to continually explain the convention of the logic system in use.

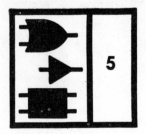

# 5     Basic Digital Circuits

Thus far, various types of arithmetic and algebra have been described without considering the actual circuit components that might be used to implement these functions. This chapter presents some of the basic circuits used in digital equipment. Typically, the circuits described in this chapter are found in modified form in many logic families. It is the purpose of this chapter merely to introduce these circuits, so that a connection can be made between the more abstract concepts of arithmetic and algebraic operations and the more familiar principles of inversion and digital switching.

For convenience in relating circuit operation to binary 1s and 0s, the logical 1 will be assumed to be the most positive voltage and the logical 0 will be represented as the most negative voltage.

## INVERTER CIRCUIT

The inverter is probably the simplest and most basic digital circuit. Yet, inverters occur as a part of almost all other more complicated circuits. The inverter's function in a digital circuit is very simple. It performs the logical NOT operation. If the input to an inverter is a logic 1, its output is represented by a logic 0. Similarly, if the inverter input is a logic 0, then a 1 is required as an output.

Normally, an inverter designed for use in logic circuits is a saturated-mode transistor switch. That is, the transistor acts very much like a mechanical switch. In the *off* condition, the transistor is in cutoff and current flow from emitter to collector is very small. Hence, the switch is effectively open. In the *on* condition, the transistor is driven well into saturation and the transistor is essentially a short circuit between the emitter and collector terminals. Here, the "switch" may be considered closed. A typical saturated-mode inverter circuit is shown in Fig. 5-1.

Initially, assume that a ground is applied at input A, representing a logic 0. It can be seen that $R_b$ and $R_k$ form a

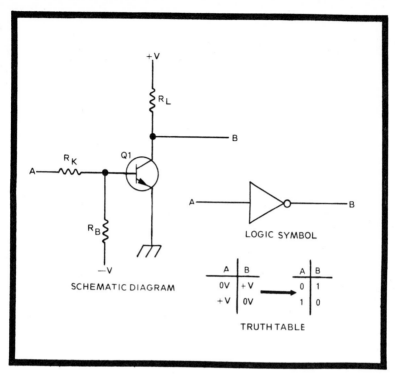

Fig. 5-1. Saturated-mode inverter circuit.

voltage divider between ground and the negative voltage, thus holding the base of npn transistor Q1 negative. This negative potential will cause Q1 to cut off, and the output at point B thus will be essentially +V. this represents the logic 1 state and is the correct operation for a NOT function as shown in the truth table.

If, on the other hand, a sufficiently large positive voltage is applied at point A, it can be seen that the base of transistor Q1 will become positive, causing the transistor to conduct heavily. The output at point B will therefore approach ground, the logic 0 level. Again, this action corresponds to the truth table for a logical NOT function.

## OR GATE

An OR gate is the name given to a digital switching circuit with performs the logical OR function. A very simple form of an OR gate is several diodes connected such that the diodes are normally biased off, representing the logic zero condition. When a logic one is applied at either diode, that diode is forward biased, thus producing a logic 1 output. This circuit, consisting of two

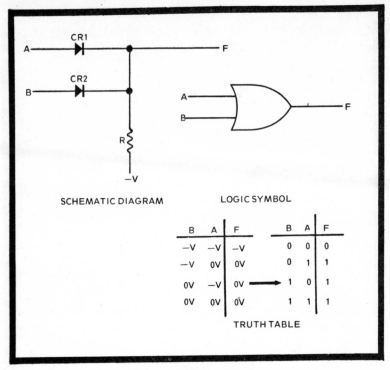

Fig. 5-2. OR-gate circuit.

diodes and a resistor returned to −V, is shown in Fig. 5-2. If point A or point B is grounded, the associated diode (CR1 or CR2) will be forward biased and a ground will be present at point F. Since ground is more positive than −V, the ground represents a logic 1 and −V represents the logical 0. The only time that output F will be at −V (logical 0) is when both A and B inputs are sufficiently negative to reverse-bias both diodes. This action corresponds exactly to the truth table defined for an OR function.

## AND GATE

The AND gate performs the logical AND function. This gate can be formed in very much the same manner as the OR gate, but with the diodes reversed and with the resistor returned to +V. An AND gate circuit of this type is shown in Fig. 5-3. The voltage which represents a logic 1 in this circuit is +V, while 0V represents a logic 0. If a ground is present at either point A or point B, one of the diodes will be forward biased, causing point F to also be at ground potential (logic 0). It is required that both points A and B be positive in order to reverse-bias diodes CR1 and

CR2. In this condition only, output F will be at +V. The truth table shows that the resultant circuit operation is identically that required for an AND function.

## FLIP-FLOP

In addition to the NOT, OR, and AND logic functions, a fourth basic circuit—called a flip-flop—is required in many digital circuits. The flip-flop is a bistable circuit which can exist in either of two stable states indefinitely and which can be made to change its state by means of some external signal. The most important use of this property is a flip-flop's ability to "store" binary information; in effect, since the flip-flop's output remains at either a 0 or a 1, depending on the last input signal, the flip-flop can be said to "remember." Another name for the flip-flop is "bistable multivibrator."

For purposes of explanation, assume that Q1 in Fig. 5-4 is initially conducting, thus causing point Q to be near ground potential. The base of Q2 will be slightly negative; so Q2 can be considered cut off, resulting in a large positive voltage at point Q̄.

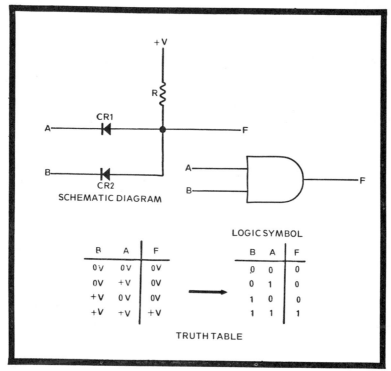

Fig. 5-3. AND-gate circuit.

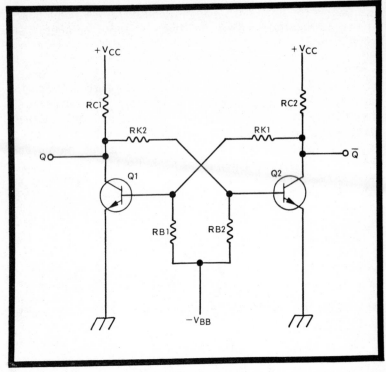

**Fig. 5-4. Flip-flop circuit.**

The resultant positive voltage at the base of Q1 will further tend to drive Q1 towards saturation, and it can be seen that the stable state for the initial conditions is Q = 0V, and Q̄ = + V.

Now assume that for a brief time a ground is placed at point Q̄ The voltage at the base of Q1 will immediately drop, cutting off Q1 and causing a large positive voltage at point Q. A positive voltage applied to the base of Q2 will cause saturation in that transistor and point Q̄ will tend to remain at ground potential. Even if the ground is no longer present at point Q̄ it can be seen that the flip-flop will remain in its new stable state, which is Q = + V and Q̄ = 0V. This property (of being capable of triggering to either state and remaining there after the triggering signal is removed) accounts for the storage property of the flip-flop.

**ONE-SHOT**

As previously described, the flip-flop has two stable states and can remain in either one indefinitely. The one-shot, however, has only one stable state and another state called a quasi-stable state. In operation, the one-shot remains in its stable state until a triggering signal is received. Upon receipt of the triggering signal, the one-shot changes to the quasi-stable state for a fixed period of time, then returns by itself to the stable state again. Since the circuit always returns to its single stable state, it is called a one-shot or single-shot. Another name for this device is a monostable multivibrator.

The circuit shown in Fig. 5-5 will be used for purposes of illustrating the operation of a one-shot. Starting with the stable state of $Q = 0V$ and $\bar{Q} = +V$, circuit operation will be analyzed. With $Q = 0V$, the base of Q1 is held slightly negative thus keeping Q1 cut off. Point $\bar{Q}$ at $+V$ keeps both sides of capacitor C at the same voltage potential and the base of Q2 is held positive, causing Q2 to be in saturation.

Now assume that a ground is placed at point $\bar{Q}$ very briefly. The ground will immediately cause capacitor C to start charging through R, forcing the base of Q2 to go negative. This action results in a change of state to the quasi-stable condition where

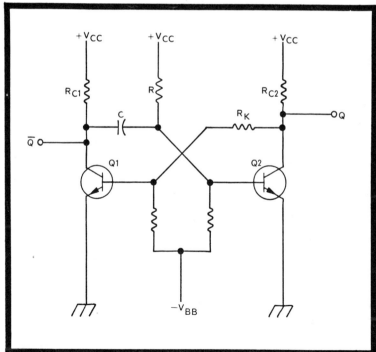

Fig. 5-5. One-shot circuit.

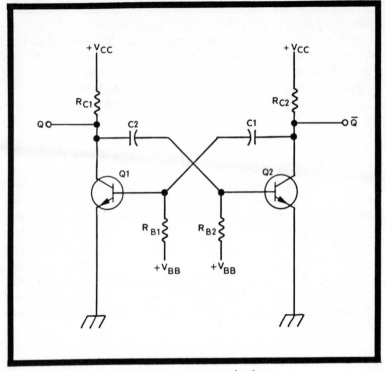

Fig. 5-6. Free-running multivibrator.

$Q = +V$ and $\bar{Q} = 0V$. However, as capacitor C continues to charge, the base voltage at Q2 will eventually rise sufficiently to cause Q2 to once again resume conduction. The length of time for this to happen is based on the time constant for R and C. The time in seconds is equal to the value of R (in megohms) times the value of C (in microfarads). When the conduction point is reached, the one-shot will return to its stable state. From the above, it can be seen that a one-shot produces an output starting with a triggering signal and ending at a time based on the RC time constant of the circuit.

## FREE-RUNNING MULTIVIBRATOR

The free-running or *astable* multivibrator does not have a stable state; instead, it continually switches between one state and the other. The time that it stays in each state is dependent upon the RC time constants of the individual resistors and capacitors used to make up the multivibrator. The most common use for a free-running multivibrator is as an oscillator which generates rectangular timing pulses for use throughout a digital unit.

A basic free-running multivibrator circuit is shown in Fig. 5-6. Assume that initially Q1 is switched to the **on** state and Q2 has just switched to its **off** state. Capacitor C1 will immediately present a large positive potential at the base of Q1, holding it in saturation, while capacitor C2 couples essentially 0V to the base of Q2. However, as time goes on, the two capacitors charge, the base of Q1 will tend to become more positive, and the base of Q2 will simultaneously become less positive. Hence, at some point, Q1 will start to conduct less while Q2 begins conducting more.

This effect is regenerative, because the changes in conduction are reflected at the two collectors, whose voltages are in turn coupled back to the bases of the opposite transistors. As a result, the circuit will switch very rapidly to the state where Q1 is cut off and Q2 is in saturation. At this point, the slow charging procedure is again initiated, this time in the opposite direction. As the potential at the two bases begin to change significantly, the regenerative action again causes a very rapid switching of states, back to the initial conditions. The above two cycles repeat indefinitely, with the timing of the cycles based upon the respective time constants, $R_{b1}$-C1 and $R_{b2}$-C2.

## SCHMITT TRIGGER

A Schmitt trigger is a bistable digital device whose state is dependent upon the amplitude of an analog input voltage. When the input voltage is below a predetermined threshold, the output is a logic 0. If the input rises above a second threshold, the output switches to a logic 1. The Schmitt trigger circuit is often used as a threshold detector to determine when some unknown input voltage has crossed the given threshold voltage. Another common use of a Schmitt trigger is to make square waves out of sine waves.

A Schmitt trigger circuit is shown in Fig. 5-7. To understand operation of the circuit, assume that initially the input to transistor Q1 is some negative voltage such that Q1 is cut off. Then a voltage divider, consisting of resistors $R_{c1}$, $R_k$, and $R_{b2}$, is formed as shown. This voltage divider establishes a positive threshold voltage at the base of Q2, dependent upon the resistive values in the divider. This positive threshold voltage will also be present at the common emitter resistor $R_e$. However, since the two transistors share a common emitter resistor, the effect is one of regenerative feedback, where the positive emitter voltage

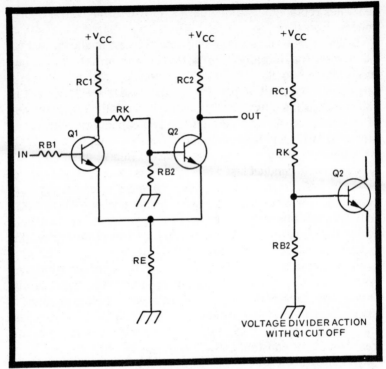

Fig. 5-7. Schmitt trigger.

serves to keep transistor Q1 in its cutoff state. Also, since the two transistors have the same emitter voltage, it should be apparent that the only way to make the circuit change state is to place a more positive voltage at the base of Q1 than the voltage which is already at the base of Q2, due to the voltage-divider action. Hence, the threshold is as previously stated, entirely dependent on the resistance values in the voltage divider.

As soon as the input voltage rises above this threshold, transistor Q1 will forward bias and the voltage at the base of Q2 will immediately drop. Through the regenerative feedback of resistor $R_e$, a new stable state will be achieved with Q1 on and Q2 off.

Now, the voltage divider still exists; however, the voltage at the junction of $R_{c1}$ and $R_k$ is somewhat lower than before, due to Q1 conducting. What this means is that even if the input voltage drops back down to the previous threshold level, the base of Q1 will still be more positive than the base of Q2 and the Schmitt trigger will not change states. What must occur is for the input to drop even lower, until the base of Q1 is indeed more negative than the base of Q2.

At this time, Q1 will cut off, causing the previous voltage-divider action; transistor Q2 will turn on. As before, the regenerative feedback of $R_e$ assures that the Schmitt trigger remains in this stable state. Note that if the input voltage is not changed, the base potential is not now sufficient to cause a state change again. Hence, the voltage difference between the two threshold voltages, called the hysteresis voltage, assures that the circuit will not oscillate at some particular voltage.

Thus, it is seen that a Schmitt trigger changes states at two separate (but related) threshold voltages. The more positive threshold must be crossed before the Schmitt trigger will switch from the logic 0 state to the logic 1 state. Similarly, the more negative threshold must be crossed to cause the Schmitt trigger to switch from the logic 1 state to the logic 0 state.

If power were applied with an input voltage between the two threshold values, the circuit would assume one of its stable states (based upon the input voltage), and then would not switch until one of the thresholds was crossed.

## INTEGRATED CIRCUITS

All of the circuits described thus far were shown schematically as consisting of individual resistors, transistors, and diodes. These components are referred to as discrete components. Another type of component is the integrated circuit. Using modern manufacturing techniques, an entire group of active circuit components can be manufactured on a single chip or semiconductor substrate. Each component could not be removed and looked at separately, but rather the entire integrated-circuit assembly can be considered as one new component.

The obvious advantages over discrete components are that integrated circuts (ICs) are much smaller than their discrete counterparts, they are almost as inexpensive to mass-produce as discretes, and there are fewer parts which can fail.

Basically, integrated circuits are constructed in much the same manner as individual diodes and transistors. Referring to Fig. 5-8, it is seen in A that each integrated circuit typically starts out as a single crystal chip of silicon, either positively (p) or negatively (n) doped. This crystal chip is referred to as the bulk or substrate material. Into the substrate, additional p and n

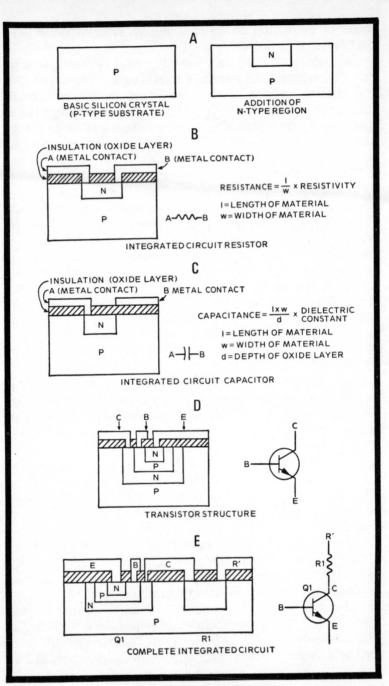

**Fig. 5-8. Integrated-circuit structure.**

dopants are added regionally through a diffusion process to form the desired components.

Sketch B shows how an integrated-circuit resistor is formed from just a single n-type region added to the p-type substrate. Metal contacts are jointed to two physically separate portions of the n-type material, using the resistance of the material to form the actual resistor. The resistor obtained is dependent primarily on the geometry of the n-type region. Since the actual resistivity of the material is usually predetermined, varying resistances are obtained by changing the length and width of the region, as required. From the given equation, it can be seen that large resistances tend to be long and narrow, whereas small resistances tend to be short and wide. An oxide layer is placed between the body of the chip and the metal contacts for insulation and circuit isolation.

To form an integrated-circuit capacitor, sketch C of the figure shows how the oxide layer then forms the dielectric material. Note that only one metal contact acutally touches the n-type material. In this case, the geometery of the n-type material, in conjunction with the thickness of the oxide layer, is the determining factor in the capacitance obtained. From the equation, it is apparent that since the depth of the oxide layer is relatively fixed, the physical size of the capacitor grows rapidly in direct proportion to the amount of capacitance required. This is an important fact to keep in mind. Inherently, integrated-circuit capacitors require fairly large areas and thus reduce the amount of circuitry which can be fabricated on a given semiconductor chip.

A simple transistor structure is shown in sketch D to illustrate the similarity in basic structure of the integrated circuit to that of the transistor. Finally, in sketch E, a composite integrated circuit made up of a transistor and a resistor is shown. Physically, the structure is much the same as if the chips from sketches B and D had been lain side by side. However, as was stated, the integrated circuit is formed on only one chip, thereby reducing the number of components, in this case from two to one. Very complex structures are made up in this same manner. Modern integrated circuits may contain several thousands of transistors and resistors on one chip.

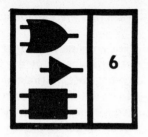

6

# Logic Families

In Chapter 5, several simple examples of logic circuits were examined and these circuits were related to specific logic functions. From a practical standpoint, logic circuits are not generally as simple as the ones shown. Further, many different types of logic families exist and, depending on the application, circuits in a particular piece of equipment may be selected from one or more of these families. Factors influencing selection of a particular logic family include (1) speed of operation, (2) noise immunity, (3) versatility, (4) power consumption, (5) size, and (6) cost, among other considerations. This chapter introduces the commonly used logic families and indicates the distinguishing features of each.

## RELAY LOGIC

Probably the simplest logic family, in terms of electronic components, is relay logic. Relay logic uses the relay windings as logic inputs and considers the resultant open or closed relay contacts as the logic 1 and logic 0 outputs.

The inverter circuit shown in Fig. 6-1 uses an input A at the relay winding to control whether or not the circuit through $R_L$ is open or closed. Initially, input A is false (at ground potential) and the relay is in its normally closed position. In this condition, the load is provided with a ground and current flows through $R_L$. When logic signal A goes true ( + V is applied), the relay opens and so does the path for current flow through $R_L$. The condition where current flows through the load is considered the logic 1 state and absence of current flow denotes the logic 0 state. From the definitions, it can be seen that the circuit of Fig. 6-1 performs the inversion (NOT) function.

It is important when dealing with relay logic to determine whether a normally open or normally closed relay is used. In the diagram, the winding is placed in the direction that the contact will move, thus indicating that when the relay is energized, the contacts will open. Hence, this example shows a normally closed relay. Resistance $R_L$ in the example typically includes the resistance of other relay windings which are being driven by the logic function. The value of $R_L$ is selected large enough to prevent the drawing of excessive current from the power supply, but small enough to assure that sufficient current is available to energize relays in the load circuit.

Examples of relay AND and relay OR circuits are shown in Fig. 6-2. In each case, it can be seen that whenever a closed circuit to ground is obtained, current can flow through $R_L$ and the output is considered to be in the logic 1 state. Examination of the truth tables for each of the circuits shows that the circuits do indeed perform the AND and OR functions as defined in Chapter 4.

A more complex example of relay logic, utilizing three input variables, is shown in Fig. 6-3. Here, the function $F = (A + B)C$ is shown with its associated truth tables. The multivariable truth tables are formed by tabulating all possible combinations of the input variables and writing down the resultant output for each combination.

Relay logic has low power dissipation when the relay is not energized and does not require a regulated power supply at all. This type of logic is ideal in a high noise environment, such as a manufacturing facility, since it is virtually immune to noise transients. Also, relays are capable of switching large currents required in some industrial applications. Several disadvantages to relay logic are: (1) its large size relative to transistors and integrated circuits, (2) its relatively slow switching speeds, and

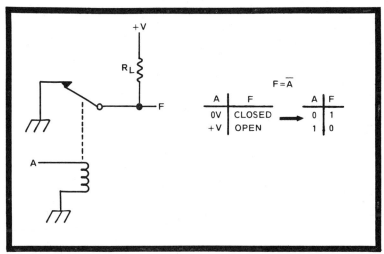

Fig. 6-1. Relay inverter.

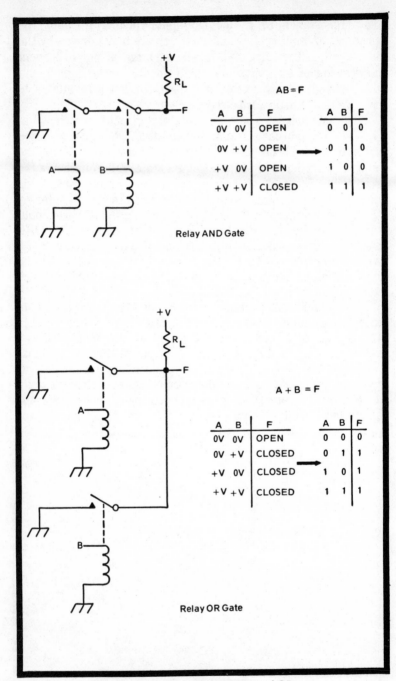

Fig. 6-2. Examples of relay AND and OR gates.

**(3)** the adverse effects of contact bounce when the relay first opens or closes.

### DIODE LOGIC

Diode AND and OR gates were described in the section on basic logic functions in Chapter 5. Therefore, only a brief review of diode logic will be included here.

A diode provides a very low impedance when biased in the forward direction and a very high impedance when biased in the reverse direction. This two-level action is similar to the opening and closing of relay contacts just described. Compared to relay circuits, diodes can operate at lower voltages and have lower overall power dissipation. Furthermore, diodes have very fast switching times and are physically quite small. The main

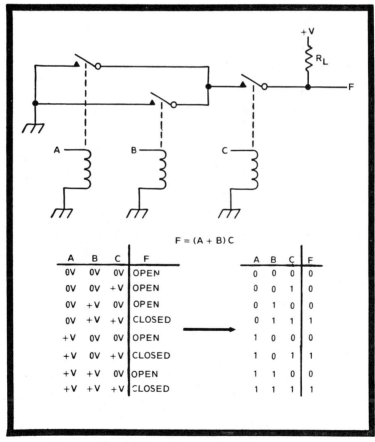

$$F = (A + B) C$$

| A | B | C | F |
|----|----|----|--------|
| 0V | 0V | 0V | OPEN |
| 0V | 0V | +V | OPEN |
| 0V | +V | 0V | OPEN |
| 0V | +V | +V | CLOSED |
| +V | 0V | 0V | OPEN |
| +V | 0V | +V | CLOSED |
| +V | +V | 0V | OPEN |
| +V | +V | +V | CLOSED |

| A | B | C | F |
|---|---|---|---|
| 0 | 0 | 0 | 0 |
| 0 | 0 | 1 | 0 |
| 0 | 1 | 0 | 0 |
| 0 | 1 | 1 | 1 |
| 1 | 0 | 0 | 0 |
| 1 | 0 | 1 | 1 |
| 1 | 1 | 0 | 0 |
| 1 | 1 | 1 | 1 |

Fig. 6-3. Complex relay logic function.

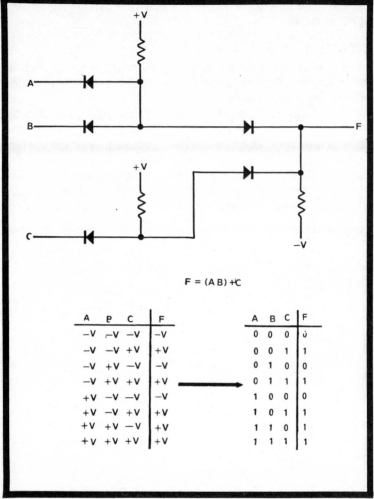

Fig. 6-4. Multilevel diode logic.

disadvantages to diode logic are: (1) the impedance of a diode circuit is quite sensitive to loading, thus providing very poor drive capability, and (2) each series diode required produces a voltage drop and a resultant shift in the typically low output voltage levels. Thus, diode logic does not lend itself well to series logic functions, and transistor power amplifiers must be used frequently to provide drive power and voltage level restoration. A multilevel diode logic circuit is shown in Fig. 6-4. In this figure, there are two levels of series diode drop. Assuming silicon diodes are used, each diode drops approximately 700 millivolts (0.7V),

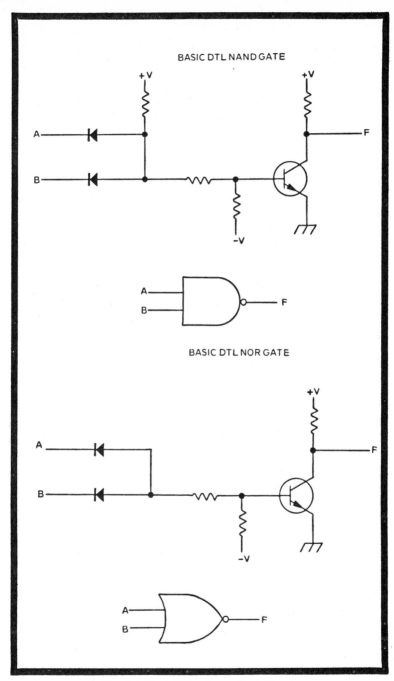

BASIC DTL NAND GATE

BASIC DTL NOR GATE

Fig. 6-5. Basic DTL gate configurations.

making a total drop in the circuit of 1.4V. Obviously, it does not take many levels of series diode logic to result in a significant voltage drop.

## DIODE—TRANSISTOR LOGIC

The basic form of diode—transistor logic (DTL) utilizes a diode AND gate followed by a transistor inverter to form a NAND gate as shown in Fig. 6-5. A DTL NOR gate is formed in a similar manner, by adding a transistor inverter to a diode OR gate. This logic family provides moderately high-speed operation and good fanout (drive) capability. Fabrication of DTL gates can be from individual transistors and diodes or in the form of integrated circuits. one of the main disadvantages to the DTL circuits shown is that both configurations require the use of positive and negative voltage power supplies for operation. A modified version of a DTL NAND gate which is more suitable for integrated-circuit applications is shown in Fig. 6-6. Here, resistors R1 and R2 form a bias feedback network which provides additional stability and temperature range. Furthermore, by adding a second transistor, increased drive capability is obtained and the requirement for a negative-voltage power supply is eliminated. The DTL family is probably the easiest to use and is also the cheapest form of logic available for medium-speed applications.

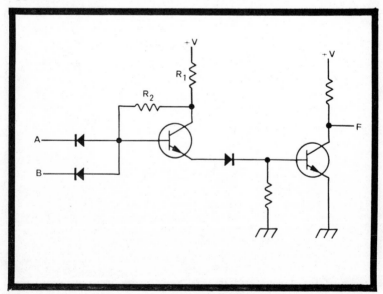

Fig. 6-6. Modified DTL NAND gate.

## DIRECT-COUPLED TRANSISTOR LOGIC

Logic gates can be made by direct interconnection of transistors. This type of logic is called direct-coupled transistor logic (DCTL) and was one of the first types of logic made into integrated circuits. Typical NAND and NOR gates using DCTL circuits are shown in Fig. 6-7. The DCTL gate is simple to make with few parts and is easy to produce in integrated-circuit form. Also, DCTL requires only a single low-voltage power supply.

Typically, DCTL circuits provide a voltage swing on the order of several volts when operating from a $+$ 3V power supply. Because of the lack of a turnoff bias for the transistor and because of the small voltage swings, the noise immunity of DCTL is poor. In their off states, the transistors in DCTL circuits operate very near the edge of conduction. For this reason, very good grounding is required; otherwise, locally generated noise is apt to cause spurious outputs which can trigger subsequent logic stages.

Another problem with DCTL is current "hogging." Current hogging results when the bases of two or more transistors are

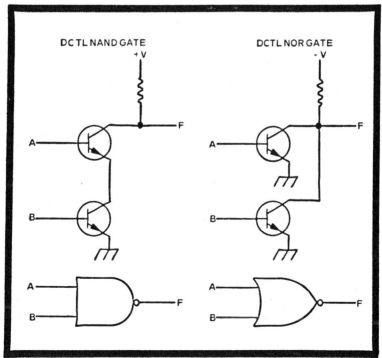

Fig. 6-7. DCTL gate configurations.

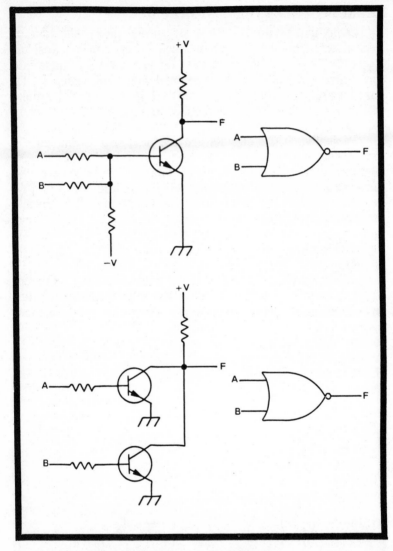

Fig. 6-8. Two versions of an RTL NOR gate.

driven directly from the collector of a single driver stage. If one transistor happens to turn on earlier than the others, the output transistor may be clamped to a value that is insufficient to turn on the other transistors. Because of this current hogging characteristic, the drive capability of DCTL is quite limited. Also, in order to minimize current hogging, transistors must be carefully selected to have very nearly identical turnon voltages.

## RESISTOR—TRANSISTOR LOGIC

Resistor—transistor logic (RTL) utilizes only resistors and transistors to make up logic gates. The RTL gates can be configured several different ways, as shown in Fig. 6-8. One configuration is similar to DTL in that the resistors peform the logic functions while the transistors serve as inverters. Another configuration is similar to DCTL but with base resistors added to eliminate current hogging.

Because of the few components required, RTL circuits are simple and reliable. Also, these circuits have relatively low power consumption and are inexpensive. Disadvantages of the RTL family are: (1) their slow switching speeds due to being driven deeply into saturation, and (2) their poor drive capability compared to other logic families.

## RESISTOR—CAPACITOR—TRANSISTOR LOGIC

The RTL transistor switching circuits just described suffer from slow switching speeds due to their tendency to draw excessive base current. This base current drives the transistor into the saturation region, resulting in an accumulation of stored charge. The time to remove the stored charge as the transistor attempts to change states is called storage delay time. One method of reducing storage delay time is to add a speedup capacitor in parallel with the base resistor, as shown in Fig. 6-9. The speedup capacitor stores the charge instead of the transistor, thus allowing the transistor to switch out of saturation more rapidly. When a resistor and capacitor are used in this manner, the resultant logic circuit is called resistor—capacitor—transistor logic (RCTL).

One disadvantage to RCTL is that the addition of the capacitor in the base circuit makes the transistor highly susceptible to noise spikes which will be coupled directly to the base of the transistor. Also, capacitors require relatively large areas in integrated circuits; thus RCTL is not the most convenient or popular form of logic for integrated-circuit fabrication.

## TRANSISTOR—TRANSISTOR LOGIC

With the advent of integrated-circuit technology, it became feasible to fabricate multiple-emitter transistors. These transistors take the place of diodes in DTL to form a new kind of logic family referred to as transistor—transistor logic (TTL or T²L). The TTL family combines high-speed operation with single-power-supply capability to make it one of the most popular

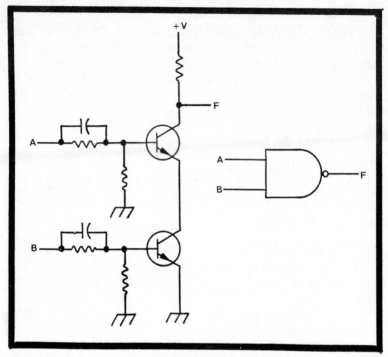

Fig. 6-9. RCTL NAND gate.

types of logic. Several configurations of the TTL NAND gate are shown in Fig. 6-10.

In the basic TTL NAND gate configuration, a low potential at either emitter will cause the emitter – base junction of transistor Q1 to be forward-biased, presenting a ground at the base of transistor Q2. Since Q2 is a standard inverter, its output will be nearly +V when ground is present at the base. When both emitters are near +V (input = high), transistor Q1 is cut off and a positive voltage is present at the base of Q2, causing Q2 to conduct. Since Q2 is conducting heavily, +V will be dropped across R2 and the output at F will be 0V.

While the basic TTL circuit is quite useful, the noise immunity of the configuration is not as good as DTL circuit, because of its inherently higher-speed switching characteristics. Also, when the output transistor is on, it represents a low-impedance current source to other circuits. However, when the output is switched off, the output reverts to a high-impedance voltage source. This high-impedance state is particularly vulnerable to pickup of noise transients.

To overcome these problems, a totem-pole output configuration is normally used. Basically, the totem-pole output provides a low-impedance output whether the output state is a logic 1 or a logic 0. This configuration results in very high-speed operation with good noise immunity. A typical TTL operating speed is 1 megahertz with propagation delays of approximately 20 nanoseconds.

One disadvantage to the use of the totem-pole output stage is the loss of the ability to connect outputs together. This is often done in DTL and RTL circuits as a means of performing

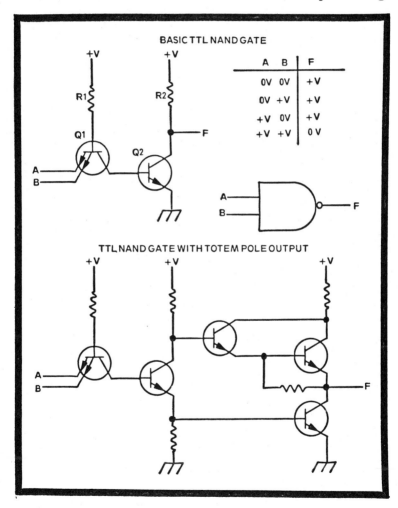

Fig. 6-10. TTL NAND gate configurations.

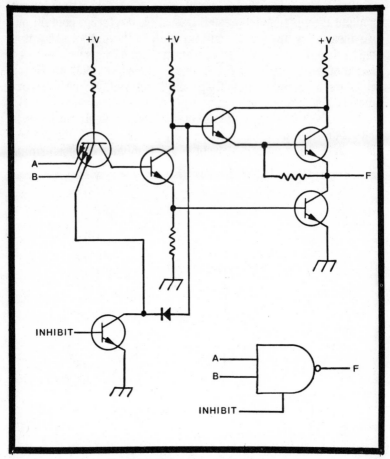

Fig. 6-11. Three-state TTL NAND gate.

additional gating and is referred to as collector ORing or "wired OR." If two totem-pole gates were to be wired together, as soon as one of the gates was in the logic 1 state and the other in the logic 0 state, there would be a low-impedance path from the supply voltage to ground, thus inhibiting normal circuit operation and permitting one or both of the output transistors to draw excessive current. Therefore, when it is desired to connect one or more outputs together, two alternatives are available, as described below.

The simplest method is through the use of open collector gates. These gates are simply the basic TTL NAND configuration with resistor R2 not provided. One common collector resistor is connected externally and as many lines as desired connected together. The disadvantage to this scheme is of course loss of the totem-pole configuration with its resultant loss of speed and drive capability.

The second approach is through the use of three-state TTL gates. The term "three-state" refers to the fact that there are three possible conditions the output can assume. These are: (1) a low-impedance logic 0 state, (2) a low-impedance logic 1 state, and (3) a high-impedance state where the output is considered essentially open. As can be seen from Fig. 6-11, when the inhibit input is a logic zero (low), the extra input emitter is in the high state and the circuit acts identically to the standard totem-pole configuration. However, if the inhibit input is set high, both totem-pole output transistors are driven to their cutoff states and the output of the circuit acts as an open circuit to other similar gates which may be connected on the same line. The only precaution which must be taken with this circuit is that only one three-state gate can be enabled at a time; otherwise, the same effects would occur as with a standard totem-pole output.

## EMITTER-COUPLED LOGIC

Thus far, all of the transistor logic circuits described have been classified as saturated-mode switching circuits. That is, the transistors are turned on by driving them into saturation. However, whenever transistors must be switched quickly out of saturation, storage delay time becomes a limiting factor in determining operating speeds.

Storage delay time can be eliminated by operating transistors in a current mode as shown in the simplified circuit of Fig. 6-12. Logic which operates in this mode is sometimes called current-mode logic (CML).

When transistor Q1 in the figure is cut off, diode CR1 is forward biased and the emitter of Q1 is held at approximately $-0.7V$. A negative voltage is required at the base of Q1 if cutoff is to be maintained. If a slightly positive voltage is applied to Q1, turning it on, diode CR1 becomes reverse biased and current flows through the transistor and through $R_E$. provided that $V_{ee}$ is much larger than the positive voltage applies at the base of Q1, the current through $R_E$ is essentially the same as when diode CR1 was conducting.

It can be seen from the above that a constant current is maintained in the emitter resistor and that the transistor never goes into saturation. Hence, the current mode of operation allows very fast switching times by eliminating saturated-mode operation.

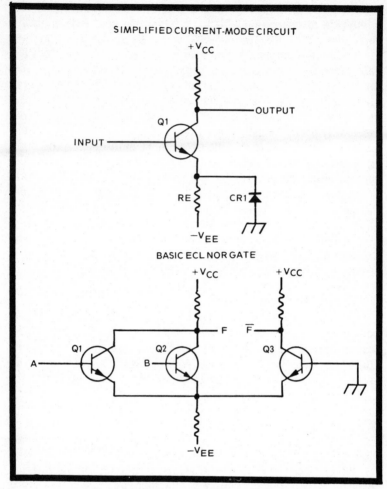

Fig. 6-12. Emitter-coupled logic schematics.

A handier form of the current-mode circuit is shown directly below the simplified circuit. A transistor has been added in place of the diode and two input transistors are connected in parallel to permit logic gating. The two-transistor combination of Q2 and Q3 is a differential amplifier pair which provides complementary outputs at the two collectors. Because of the common emitter resistor used in the differential amplifier, the circuit is referred to as emitter-coupled; logic which utilizes this configuration is often referred to as emitter-coupled logic (ECL). Thus, CML and ECL are two different names for the same type of circuit.

In the schematic shown, if both inputs A and B are slightly negative, transistors Q1 and Q2 will be cut off and the emitter—base junction of Q3 will become forward biased, causing Q3 to conduct. If either of the inputs goes slightly positive, the associated transistor (Q1 or Q2) will conduct, cutting off transistor Q3.

One problem with ECL is that it is a low-level logic. This factor makes ECL quite noise-susceptible, and great care must be taken to reduce noise spikes and ground noise. However, the extra care required may well be worthwhile where the high operating speeds of ECL are necessary. Disadvantages to ECL are: (1) multiple power supplies are required, and (2) ECL does not readily interface with other logic families without special buffering.

### Metal-Oxide Semiconductor (MOS) Logic

MOS logic utilizes field-effect transistors (FETs) in place of conventional bipolar transistor to form gates and delay elements. From the basic definition of the FET, it will be recalled that the field-effect transistor operates by using an electric field to control current flow in a conducting channel. The current in the channel is derived from the flow of majority carriers; holes for $p$-channel MOS, or electrons for $n$-channel MOS.

Early versions of the insulated-gate field-effect transistor used a metal-oxide-semiconductor layered construction, similar to the one shown in Fig. 6-13. The metal portion was aluminum, used to form the gate, which was separated from a silicon semiconductor channel by an insulating layer of silicon dioxide. Today, however, the term is not limited to this specific constuction and refers to any insulated-gate field-effect transistor. There are two basic MOS structures which should be understood. These are depletion-mode MOS devices, and enhancement-mode MOS devices. Both are generally termed "MOSFETs."

The depletion-mode MOS structure is that shown in Fig. 6-13. A physical channel of $n$ or $p$ material is diffused into a lightly doped substrate material. At one end is a drain terminal. The flow of current between source and drain is modulated by a voltage applied to a gate terminal which is insulated from the source and drain. In the illustration, when the gate voltage is at ground potential, the insulation acts as a capacitor dielectric to cause the $n$ material to be basically uncharged. As a result, electrons flow freely through the channel. However, when the

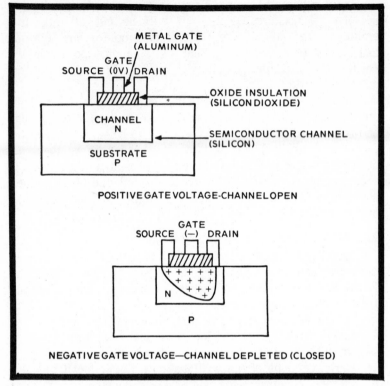

Fig. 6-13. Depletion-mode MOS device.

gate voltage is made negative, the capacitive effect of the gate plus insulation is to induce a positive potential into the channel, causing depletion of electrons in the $n$ material. As a result, the channel is "pinched off" so that current flow is reduced or cut off. The positive charge in the $n$ channel is effectively a depletion of electrons, the majority carriers, hence the name depletion-mode MOS.

The enhancement-mode MOS device is shown in Fig. 6-14. This device does not actually have a physical channel, but instead has two $n$-type wells at either end of the gate structure. In this case, the gate can be modulated to induce a channel between the two $n$-type wells. Since this can be considered an enhancement of electrons in the channel, the device is referred to as enhancement-mode MOS. Assuming the quiescent condition where the gate is 0V, the $p$-type substrate material does not permit current flow between the two $n$-type wells, and the channel is effectively closed. But when a positive gate voltage is applied, the gate capacitive effect induces a negative potential in the p-type substrate, thereby inducing a channel and supplying electrons for current flow between the two wells.

The most obvious difference between the two devices; the enhancement-mode, and depletion-mode, is that in the quiescent condition (no gate voltage), current flows in the depletion-mode device but not in the enhancement-mode device. For this reason, enhancement-mode MOS transistors are used in most applications. Both examples shown were for n-channel MOS devices. However, p-channel MOS operates in a similar manner except that the p-channel MOS requires opposite polarity gate voltages for the same conditions.

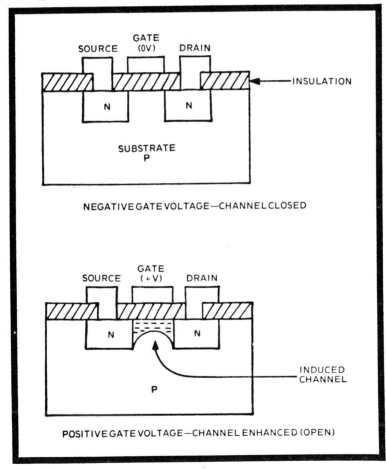

Fig. 6-14. Enhancement-mode MOS device.

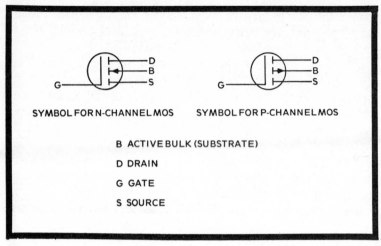

**Fig. 6-15. Schematic symbols for MOS devices.**

The operation of MOS logic gates can easily be explained from the schematic diagrams for either $p$ or $n$ MOS circuits. Since the main differences are in the polarity of the supply voltages, the following description will be provided only for $p$-channel MOS circuits.

A typical MOS inverter circuit is shown in Fig. 6-16. One of the most apparent differences between this circuit and the circuit of a bipolar transistor inverter is the substitution of a transistor in place of a resistor as a load.

Basically, MOS devices are integrated circuits which are made up on silicon chips. A resistor on one of these chips requires a fairly large area, while a transistor can be fabricated in a small area. For this reason, a transistor normally serves as the load. The load transistor is held conducting at all times by maintaining its gate voltage ($-V_G$) slightly more negative than the drain voltage ($-V_D$).

When the input signal is at 0V, the MOS transistor is off and the output is approximately $-V_G$. provided that the gate voltage is close to the drain voltage, the voltage drop across the load transistor is very small. When the input signal goes negative, the input transistor is turned on and the output falls to ground potential. In this case, approximately $V_D$ is dropped across the load transistor.

Operation of NAND and NOR gates is quite similar to the basic inverter operation. The NAND gate has two transistors in parallel, while the NOR gate has two transistors in series. It should be remembered that since *p*-channel MOS uses negative supply voltages, a logic 1 is often assigned as the negative voltage and a logic 0 as ground.

When using negative-true logic in this manner, the gates just described serve opposite functions from those noted. That is, the NAND gate in negative-true logic becomes a NOR gate, and vice versa.

The main advantages to MOS logic are its high noise immunity due to large logic voltage swings and, as a result, the

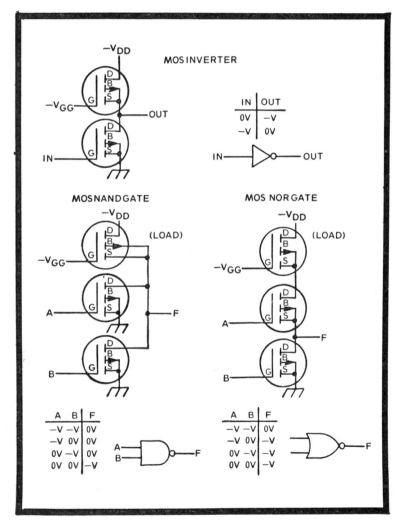

Fig. 6-16. Examples of p-MOS logic gates.

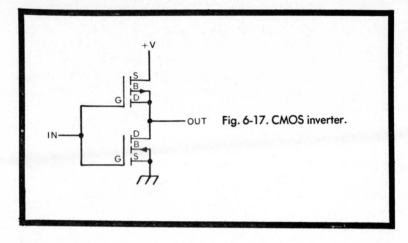

Fig. 6-17. CMOS inverter.

ability to use unregulated power supplies which are considerably cheaper than regulated supplies. Another significant advantage to MOS logic is that very high packaging densities can be obtained compared to bipolar speeds than other types of logic and that it is difficult to interface MOS with other logic families.

### Complementary MOS

Complementary MOS (CMOS) circuits utilize both $n$-MOS and $p$-MOS transistors on a single chip to form circuits which have very low power dissipation. Further, CMOS circuits are typically operated from a single power supply instead of two supplies as are used with regular MOS. A typical CMOS inverter circuit is shown in Fig. 6-17. The inverter consists of an $n$-channel MOS transistor in series with a $p$-channel MOS transistor. When the input to the inverter is 0V, the $n$-MOS transistor is off and the $p$-MOS transistor is on. Thus, the output of the inverter is + V for a 0V input. When the input switches to + V, the $n$-MOS transistor turns on and the $p$-MOS transistor switches off. In this case, the output goes to 0V and it can be seen that the output in each case is inverted from the input. No matter which state the input assumes, one MOS transistor is on and the other is off. As a result, steady-state power consumption for CMOS circuits is very low and the only time that large amounts of current are drawn is during actual switching operations, the transition from off to on.

A CMOS NOR gate is formed by connecting two $n$-MOS transistors in parallel, and two $p$-MOS transistors in series, as shown in Fig. 6-18. When both of the inputs, A and B, are at 0V; the two $p$-MOS transistors are on and the two $n$-MOS transistors are off. Thus, + V is available at the output, F. When either

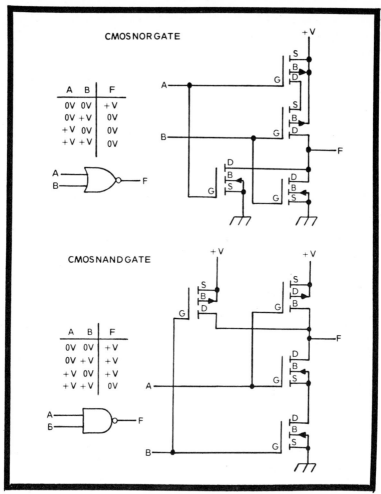

Fig. 6-18. Typical CMOS gate structures.

input changes to +V, the associated $n$-MOS transistor turns on, completing a path to ground; then, the $p$-MOS transistor opens, disconnecting +V from the output. The resultant output is therefore 0V for any +V input.

Similarly, a CMOS NAND gate is formed by connecting two $n$-MOS transistors in series and two $p$-MOS transistors in parallel. If either input goes to 0V, the associated $p$-MOS transistor is turned on and the series path to ground is opened by the $n$-MOS transistor turning off. Thus, the output goes to +V. When both inputs are at +V, the $p$-MOS are turned on, connecting a ground to the output.

# Sequential Logic

In general, logic functions are performed with a mix consisting of two distinct types of logic circuits. These are (1) combinational circuits which include AND, OR, and NOT functions and, (2) sequential circuits which are made up with various types of storage or delay devices. Typical delay devices which make up sequential logic circuits are the subject for this chapter.

Sequential logic is so named because logic operations performed with these devices occur in a definite sequence. For example, the results of one operation such as binary addition may be stored at some initial time, say time A, in a group of flip-flops. At a later time, say time B, the outputs of these flip-flops may be utilized to perform another logic operation such as multiplication. In this manner, a number of operations may be performed in sequence, each operation depending upon the results of previous operations. A digital computer is an example of a group of circuits which operates in this manner.

## FLIP-FLOPS

The basic delay device used in most logic families is the flip-flop. A simplified schematic of a flip-flop was shown in Chapter 5 and the storage or delay capability of the flip-flop was also noted at that time. There are several types of flip-flops which will be encountered and each of these has characteristics which distinguish it from other types. The following descriptions outline the unique characteristics of each type of flip-flop. (A truth table and specific logic symbols are given in each case.) Timing diagrams are included as an aid to understanding the sequential nature of the circuits. As with normal waveforms associated with analog circuits, the timing diagrams are used to present a time-versus-voltage history for the various circuit inputs and outputs.

There are several unique terms used to describe flip-flop operation. There are two complementary outputs from a flip-flop, and these are normally labeled $Q$ and $\bar{Q}$. When one output is in the logic 1 state, the other output is always a logic 0. If the flip-flop changes states, then both Q and Q̄ change.

A flip-flop is considered to be **set** when Q = 1 and Q̄ = 0. Conversely, the flip-flop is **reset** when Q = 0 and Q̄ = 1. Thus, the process of causing the flip-flop to go to the Q = 1, Q̄ = 0 state is called *setting* the flip-flop; and causing the flip-flop to go to the Q = 0, Q̄ = 1 state is called *resetting* the flip-flop (this action is also referred to as "*clearing*").

### NAND Gate SR Flip-Flop

A simple flip-flop can be made from two cross-connected NAND gates as shown in Fig. 7-1. If the two inputs are labeled S and R (for set and reset), then the flip-flop can be called a set–reset or SR flip-flop. Referring to the figure, it can be seen that if the S input goes to a logic 0, the flip-flop will go to its set state (Q = 1) and will remain there until reset. When the R input goes to a logic 0, the flip-flop will go to the reset state and stay there until it is set. Thus, an SR flip-flop changes states upon

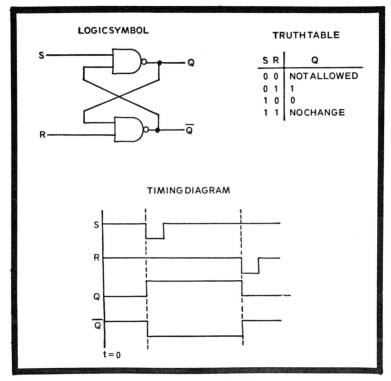

Fig. 7-1. NAND gate SR flip-flop.

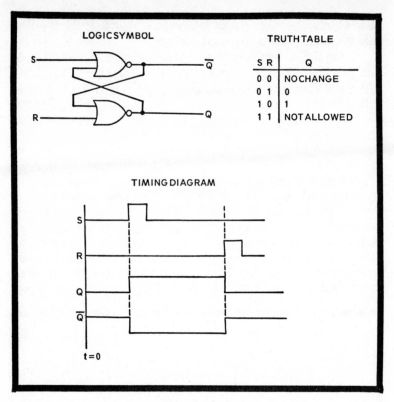

**Fig. 7-2. NOR gate SR flip-flop.**

sensing a change in state at the S or R inputs and stores the results of the change until the opposite input is activitated.

There are two special conditions of interest for this flip-flop. First, notice that the condition S = 0, R = 0 is labeled as being "not allowed." This means that the operation of the circuits which send inputs to S and R must be restricted such that they never go to a logic 0 at the same time. If this condition were to occur, it can be seen from examination of a basic NAND gate truth table, that both Q and $\overline{Q}$ would go to a logic 1 at the same time. Of course, this is not the defined operation of a flip-flop. Secondly, observe that the condition S = 1, R = 1 is the quiescent state for this flip-flop. As long as both inputs remain at a logic 1, no change in the state of the flip-flop is possible.

### NOR Gate SR Flip-Flop

The circuit shown in Fig. 7-2 is a flip-flop which is configured by cross-connecting NOR gates in a manner which is very similar to the NAND gate version previously described. The main difference between the two is that the activity changes at the S and R inputs are logic 1s instead of logic 0s. Therefore, for the NOR gate configuration, the quiescent state is S = 0, R = 0, such that no change will occur and the *not allowed* state is S = 1, R = 1. Since this state is not allowed, the output for this condition is often referred to as undefined or *don't care*. This state becomes important when considering logic minimization techniques to be described in a later chapter.

### Clocked Flip-Flops

To describe the operation of clocked flip-flops, the concept of a synchronous system is introduced. A synchronous system is controlled by a master oscillator and wave-shaping circuit which produces a set of *clock* pulses. These clock pulses occur at some fixed interval (for example, every 10 microseconds) and all logic state changes are synchronized to occur at the time when the clock pulse occurs. Clocked flip-flops are the means by which this synchronization is maintained.

The truth table for a clocked flip-flop is defined in terms of an output (Q) at the next clock pulse, if the state of the output is known at the present clock pulse. Thus, if the output of the flip-flop for clock pulse n is some arbitrary state Q—that is, either one or zero—then at clock pulse n + 1 the new output is defined. The new output might be state Q (no change from the previous clock period), state $\bar{Q}$ (the opposite from what it was) ; it could be known to be a logic 1, a logic 0, or perhaps an undefined state. As previously noted, an undefined output state merely denotes a *not allowed* combination of inputs.

### Clocked SR Flip-Flop

Basically, a clocked SR flip-flop operates in the same manner as the NAND and NOR gate versions with the exception that the clock pulse controls the times at which state changes can happen. Referring to the timing diagram of Fig. 7-3, it is seen that the Q and $\bar{Q}$ outputs do not respond directly to the S and R inputs; rather, they await the next clock pulse before changing states. For this flip-flop, if the output is in state Q when clock pulse *n* occurs, it will remain in that state for clock pulse n + 1 provided that inputs S and R are both zero. If either S or R goes to the logic 1 state, then the flip-flop will assume the appropriate state upon occurrence of the next clock pulse. Finally, as with the other SR flip-flops, there is a **not allowed** condition, namely S = 1, R = 1, for which the output is undefined. This condition is represented by a question mark in the truth table.

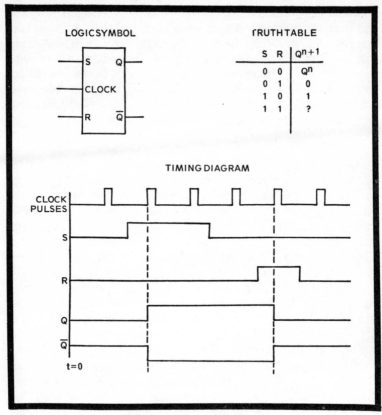

LOGIC SYMBOL

TRUTH TABLE

| S | R | $Q^{n+1}$ |
|---|---|-----------|
| 0 | 0 | $Q^n$ |
| 0 | 1 | 0 |
| 1 | 0 | 1 |
| 1 | 1 | ? |

TIMING DIAGRAM

Fig. 7-3. Clocked SR flip-flop.

## T Flip-Flop

A commonly required function is performed by the T flip-flop shown in Fig. 7-4. This flip-flop has the ability to change states (toggle) each time a clock pulse occurs. The T flip-flop has one input, T, which controls whether or not the toggling action will occur. When T is set to a logic 1, the flip-flop toggles, and when T is a logic 0, the flip-flop remains in its current state. This ability to toggle with each clock pulse is the basis for digital counters and frequency dividers.

Referring to timing diagram A, it is seen that the result of changing state each time is, in effect, a divide-by-two of the basic clock frequency. This is shown even more graphically in timing diagram B, where a wide clock pulse is used; it can be observed that the output (Q) is indeed a square wave at one-half the frequency of the clock pulse.

Careful examination of the states of the two waveforms, clock vs outputs, shows that while T is a logic 1, the waveforms go through a binary count sequence. That is, the states are 00, 01, 10, and 11, in that order. Here, the first digit represents the Q output states and the second digit represents the clock pulse states. In other words, the T flip-flop operates as a binary counter. If several T flip-flops are connected together in an appropriate manner, any desired count sequence or frequency division can be obtained. Various counter configurations are described in subsequent chapters.

## D Flip-Flop

If an inverter is placed at the R input to an SR flip-flop, as shown in Fig. 7-5, the result is a D or delay flip-flop. Since the S

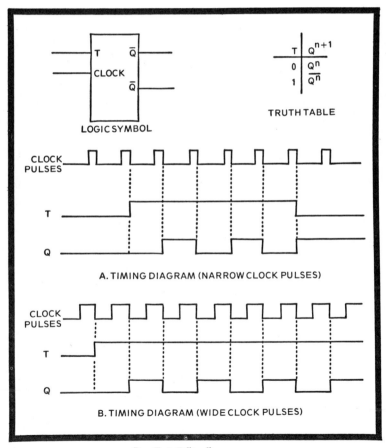

Fig. 7-4. T flip-flop.

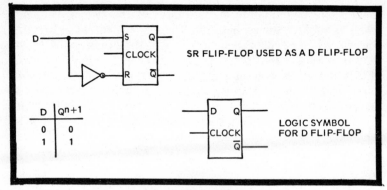

Fig. 7-5. D flip-flop.

and R inputs are always in opposite states, the flip-flop merely follows the state of whatever data is present at the D input. More importantly, the result of this action is to delay the data at the input by one clock pulse. In other words, the flip-flop can be considered to have stored the data from the time of the first clock pulse until such time as another clock pulse occurs.

However, D flip-flops are not necessarily built from SR flip-flops, and a circuit which operates as a D flip-flop may not be capable of performing as an SR flip-flop. Thus, a logic symbol as shown in the figure is used to depict a D flip-flop. An important usage of D flip-flops is in the interconnection of several flip-flops in series to form **shift registers** or in parallel to form **data storage registers**. In the shift register configuration, delays of many clock periods can be obtained. When used as storage registers, the flip-flops can store data indefinitely, or until required. Shift registers and storage registers have many applications in digital data handling circuits.

### JK Flip-Flop

Perhaps the most useful of all the flip-flop configurations is the JK flip-flop, the symbol for which is shown in Fig. 7-6. A comparison of the truth tables for the JK flip-flop with that for the clocked SR flip-flop reveals that the difference between the two is that the $J = 1$, $K = 1$, condition is allowed, whereas the $S = 1$, $R = 1$ condition is not allowed. In the JK flip-flop, the $J = 1$, $K = 1$ condition causes the flip-flop to toggle. With this one modification, the JK flip-flop is created which can be used for any of the previously described flip-flop functions. For example, it is obvious that the JK flip-flop can be used as an SR flip-flop, since the only difference between the two is a state which is not allowed for SR operation anyway. Further, if the JK inputs are tied together as shown, the truth table reduces to that of the T flip-flop previously described, and functionally the two circuits are identical. Finally, if an inverter is added, such that J is always the inverse of K, the truth table reduces to that of the D flip-flop. Again, the two circuits are functionally identical. JK flip-flops are often used as the basic flip-flop element in a system design due to their ability to be used in these various ways.

## MOS DYNAMIC STORAGE

A completely different type of storage device from flip-flops is the dynamic storage obtained from the use of MOS (metal oxide semiconductor) transmission gates in conjunction with MOS inverters. All storage with a conventional flip-flop is considered static, in that data stored in a flip-flop can be held indefinitely. The storage obtained through the MOS transmission gates, however, is considered **dynamic**, in that data can only be held for a limited time before it must be refreshed (stored again).

In essence, the MOS transmission gate is a single-pole, single-throw semiconductor switch which can be used to pass current to a capacitor and then be shut off to prevent the

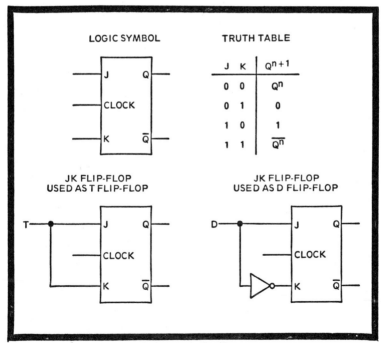

Fig. 7-6. JK flip-flop.

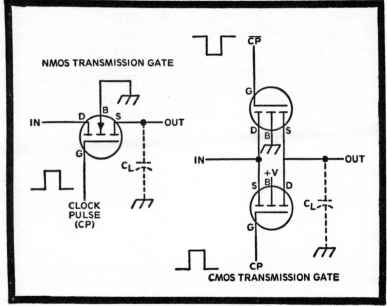

**Fig. 7-7.** MOS transmission gates.

capacitor from discharging at a later time. However, the key to the whole storage scheme lies in the fact that an actual capacitor is not required. Happily, the gate terminal of a MOS transistor looks like a capacitor with almost no leakage. Thus, the transmission gate is used to charge up the gate capacitance of another MOS transistor in an inverter stage; then, when the gate closes, the charge is stored in that stage and the stored charge does not leak off.

To illustrate how the transmission gate is employed to charge this capacitance, consider for example the $n$-MOS transmission gate shown in Fig. 7-7. Assume that the input signal is some positive voltage at the drain terminal and that the clock pulse is initially low. With the clock pulse low, the gate is off and the capacitive "load" remains uncharged. Now, when the clock goes positive, the gate turns on and the load capacitance rapidly charges. At a still later time, when the clock goes low again, the gate opens up again and the charge is now stored by the load capacitance. A CMOS (complementary MOS) transmission gate is also shown for reference; however, its operation is quite similar to that of the n-channel device.

It is apparent that a low-speed limit and a high-speed limit are both imposed on this circuit. At the low-frequency end, it must be remembered that there is some small amount of leakage which will occur. If the clock frequency is not sufficient to read out data faster than the capacitive decay rate, the proper logic levels will no longer be present when an attempt is made to read the data. A high-frequency limit exists because a finite amount of time is required to charge the gate capacitance. If the clock frequency is too fast, the capacitance will not become charged sufficiently, and again the proper logic levels will not be present.

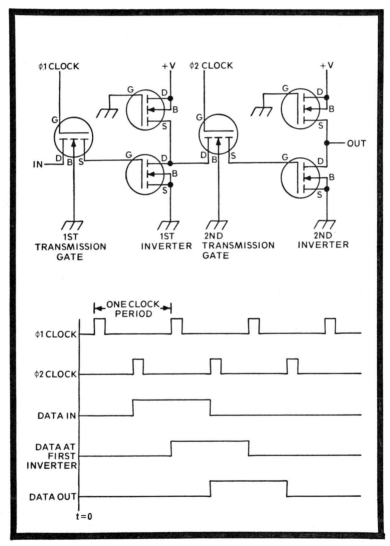

Fig. 7-8. Single stage of MOS dynamic storage.

If the transmission gate of Fig. 7-7 were simply used with an inverter, and a series connection made of a group of these circuits, there would be no storage at all. This is because at the next clock period the load capacitance would be discharged by the previous stage at the same time it was being read into the following stage.

This problem is solved by using a two-phase clocking system similar to that shown in Fig. 7-8. Here, input data is determined such that it changes states at the phase-two clock leading edge. Thus, when the phase-one clock occurs, the input data is settled to its desired state and the capacitance of the first inverter can charge up. After the phase one clock goes low, the first inverter has stored the signal until the phase-two clock pulse occurs. At this time, the first inverter charges the second inverter capacitance through the transmission gate, and the charge is now stored at the second inverter. This series performance can be repeated for as many of these two-phase stages as desired, each stage providing a delay equal to one clock period.

# Analysis of Logic Networks

When analog networks are studied, it is very often at the detailed schematic-diagram level that an understanding of the overall network function is obtained. However, digital circuits must be understood almost entirely at the functional level, where individual resistors and transistors are of little or no importance. Instead, entire logic functions such as AND, OR, NOT, etc. are considered either in diagram form, through logic equations, or via truth tables. Nevertheless, each of the available forms has drawbacks which may make it difficult to visualize the functions or to see where redundancies and problem areas exist. For this reason, some additional tools are required to aid in the analysis of logic networks. These tools include algebraic manipulation of logic equations, mapping of the logic functions, and the preparation of flow diagrams.

## ALGEBRAIC MANIPULATION

In Chapter 4, the concept of the logical AND, OR, NOT, and equality were presented, and these functions were related to symbols used in Boolean algebra. As with ordinary algebra, there are a number of rules which are associated with Boolean algebra which, if followed, permit expressing logic functions in a number of different ways. While each expression of a function yields the same result, some forms may be simpler to understand than others. Thus, it is desirable to know the various rules which are permitted and to use them to advantage whenever possible. Table 8-1 contains a listing of the basic relationships often used.

To illustrate the understanding which may be gained through algebraic manipulation, consider the truth table for a NAND function as given in Fig. 8-1. If the function is thought of in its conventional sense, the concept is that there is a logic 1 output any time the condition $A = 1$, $B = 1$ does not exist. This statement agrees with the truth table and can be expressed algebraically as $F = \overline{AB}$. However, for some people, this is "backward" thinking and they may prefer to consider the circuit

## Table 8-1. Boolean Algebra Relationships.

| | |
|---|---|
| 1. $A + A = A$ | 13. $A \cdot BC = (A \cdot B)(A \cdot C)$ |
| 2. $AA = A$ | 14. $\overline{A + B} = \overline{A}\overline{B}$ |
| 3. $A + 0 = A$ | 15. $\overline{AB} = \overline{A} + \overline{B}$ |
| 4. $A \cdot 0 = 0$ | 16. $A + AB = A$ |
| 5. $A + 1 = 1$ | 17. $A(A + B) = A$ |
| 6. $A \cdot 1 = A$ | 18. $(A + B)(A + \overline{B}) = A$ |
| 7. $A + \overline{A} = 1$ | 19. $AB + A\overline{B} = A$ |
| 8. $A\overline{A} = 0$ | 20. $A + \overline{A}B = A + B$ |
| 9. $\overline{\overline{A}} = A$ | 21. $A(\overline{A} + B) = AB$ |
| 10. $A + B = B + A$ | 22. $\overline{A}(A + B) = \overline{A}B$ |
| 11. $AB = BA$ | 23. $\overline{A} + AB = \overline{A} + B$ |
| 12. $A(B + C) = AB + AC$ | |

action in terms of a logic 1 output if either input is a logic 0. This statement also agrees with the truth table, but its algebraic expression is of the form $F = \overline{A} + \overline{B}$.

Fortunately, rule 15 of Table 8-1 states that these two equations give identical logical outputs and that both concepts of the NAND function are correct. Note also that the function may in reality be implemented with an actual NAND gate, but it might just as well be implemented with an OR gate. Both circuits have identical outputs. Finally, this example brings home an important point. That is, all logic functions are **defined** in terms of their truth tables. A given function may be manipulated algebraically, or implemented with different types of gates, but its truth table never changes—by definition!

Also shown in Fig. 8-1 is an example of the same kind of equivalency for the NOR function. In the NOR case and the NAND case, it is seen that whenever there is an AND representation for a function, there is a comparable OR representation also. This AND – OR equivalency many times helps in understanding a function whose variables always seem to be inverted from the way one would like to think of them.

An entirely different usage for algebraic manipulation is suggested by Fig. 8-2. Here, the truth table for a simple OR function is given. But assume that, for whatever reasons, the truth table is not recognized as an OR function and the equation

$F = AB + A\overline{B} + \overline{A}B$ is written, with its resultant implementation as shown. A single application of rule 19 shows that the function can be reduced to the form $F = A + \overline{A}B$, requiring one-half the gates initially used.

If rule 20 is also applied, it becomes obvious that the function is indeed an OR function and can be implemented with a single gate. In each case, the truth table was identical, while the logic equations were considerably changed. Although it is not always possible to effect a savings in logic hardware, the point to be remembered is that algebraic manipulation may permit writing a logic equation which is more easily understood.

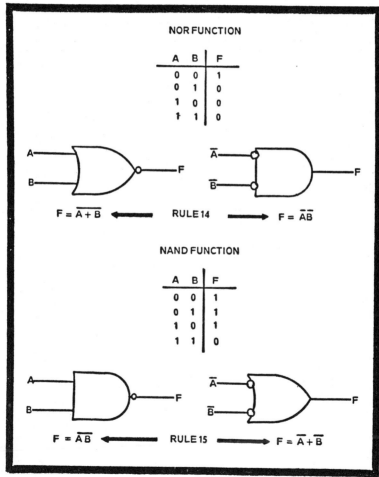

Fig. 8-1. Equivalency of logic circuits.

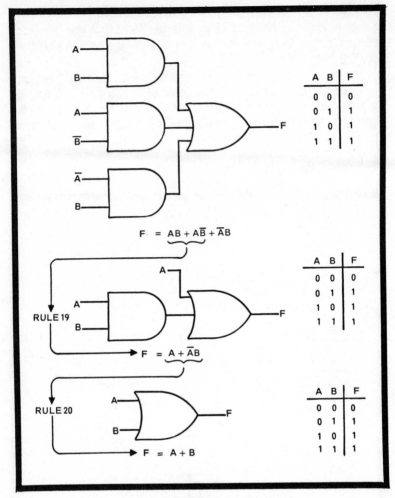

**Fig. 8-2. Simplification by algebraic manipulation.**

## KARNAUGH MAPS

Another method for examining the states of a logic function is through the use of Karnaugh maps. Karnaugh maps are representations of truth tables which have been organized into cells. Each cell represents a 1-bit change in the input variables when compared to the next adjacent cell. The truth table for a NOR function is shown in Fig. 8-3 as an example of a two-variable function which can be transferred to a Karnaugh map. In this map, a 1 has been placed in the cell corresponding to the $A\bar{B}(00)$ state, while a 0 is placed in each of the other three cells. It is

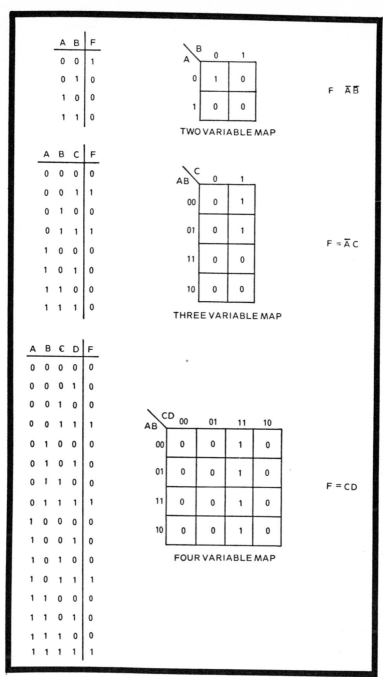

Fig. 8-3. Basic Karnaugh map configurations.

obvious from this map as well as from the truth table that one equation which will represent the function shown is $F = \overline{A}\overline{B}$.

## MULTIVARIABLE FUNCTION

A three-variable Karnaugh map can be formed from a three-variable truth table in a similar manner. Upon inspection of this truth table, it might be thought that a reasonable equation to represent the given function is $F = \overline{A}\overline{B}C + \overline{A}BC$. However, the Karnaugh map shows that an output of one is obtained no matter which state variable B assumed and that the only requirement to fulfill the function is that $F = \overline{A}C$. Of course, all this can be verified or derived by manipulation of the logic equations. As a general rule, however, algebraic manipulation is more helpful in expressing a function in some other form, while the Karnaugh map is useful for obtaining the simplest possible expression of the truth table.

A final example will illustrate the application of a Karnaugh map to a four-variable function. Here, the truth table shows that a logic 1 output is obtained for four different conditions. It might not be clear from the truth table that each of these conditions does not require a four-input gate. Reference to the Karnaugh map, however, makes it clear that all the logic 1 outputs are obtained when C and D are both a logic 1 and that the states of A and B do not matter. Thus, an equation which adequately represents this fuction is $F = CD$, which can be implemented with a single two-input NAND gate.

### Using the Karnaugh Map

Various groups of logic 1s in a Karnaugh map can be considered together and a resultant simplification of the required logic implementation obtained. Further, it should be clear that even if a truth table does not exist, one can be generated from logic equations or from a logic diagram

From the truth table, the logic 1 outputs can be transferred into a Karnaugh map. All that remains is to determine which groupings of logic 1s can be considered together to help in writing a simplified logic equation.

All the combinations of allowed groupings, called loops, are shown in Figs. 8-4 and 8-5, for three- and four-variable maps, respectively. Each loop in a map consists of a single AND function, and all the AND functions which are represented by the various loops are ORed together to obtain the final logic network. Since this method requires one AND function for each loop, a minimum number of loops should be used. Additionally, it is always desirable to draw the largest possible loops, because the larger the loop, the fewer the number of variables contained in each AND function. Translated into logic gates, the smaller the resultant logic equation, the fewer the number of gates required to implement the equation.

As an example of the use of loops to aid in writing a logic equation, consider the logic network defined in Fig. 8-6. It is not at all clear from the truth table which functions are being performed by this network. Even after the truth table is transferred to the Karnaugh map, the logic is not easy to

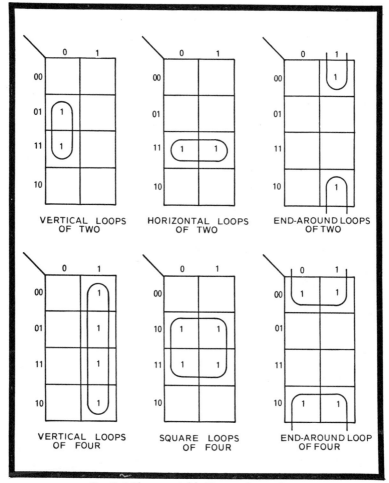

Fig. 8-4. Three-variable Karnaugh map loops.

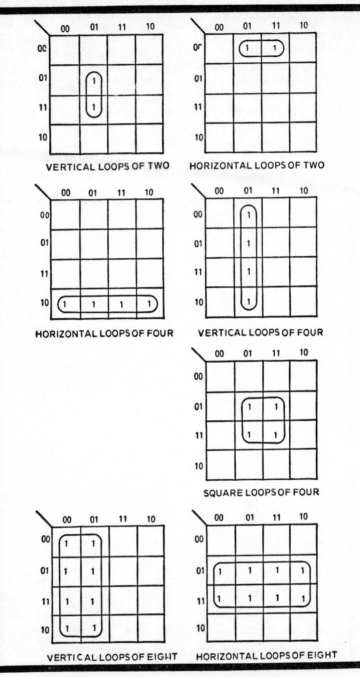

Fig. 8-5. Four-variable Karnaugh map loops.

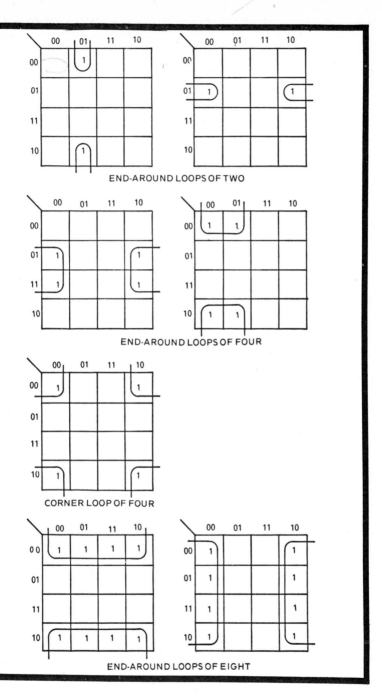

END-AROUND LOOPS OF TWO

END-AROUND LOOPS OF FOUR

CORNER LOOP OF FOUR

END-AROUND LOOPS OF EIGHT

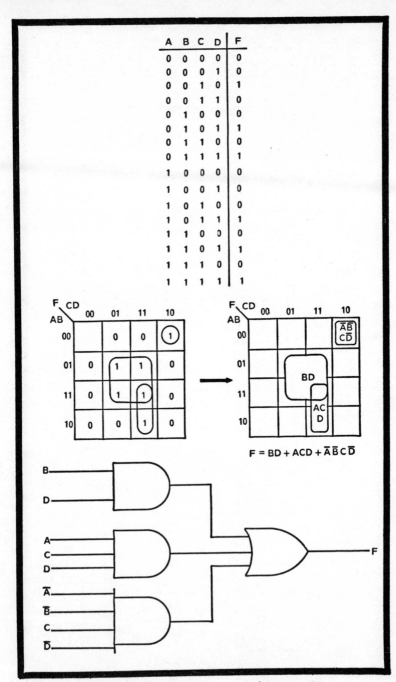

Fig. 8-6. Example use of loops to write logic equation.

visualize. However, by drawing loops as shown, a simple equation can be written which aids greatly in understanding the logic functions called for by the truth table. Further, if desired, a simple logic diagram similar to that shown can be drawn which corresponds to the simplified logic equation.

## FLOW DIAGRAMS

Sequential networks usually employ flip-flop elements which change states from clock pulse to clock pulse, as determined by logic inputs and the logical interconnection of the flip-flops. To represent the changing states of these sequential networks, a truth table arrangement similar to that shown in Fig. 8-7 is often employed. However, with the large quantity of 1s and 0s contained in such a table, the table often becomes quite confusing. A method for clarifying this type of sequential operation is the flow diagram. In such a diagram, the various combinations of flip-flop states are assigned arbitrary state numbers and the state numbers are interconnected with arrows which show the next state that the network will assume. The combination of inputs which causes each particular state change is shown directly next to the arrowhead.

Referring to the simple counter circuit of Fig. 8-7, the truth table indicates that when the control variable C is a logic 1, the network counts in a standard binary sequence. Further, when the control variable C is a logic 0, the network remains in whatever state it was last in. However, even for this simple case, the complexity of the truth table is such that the circuit action is not apparent without careful inspection. Thus, a flow diagram is used to help visualize circuit operation. The flow diagram clearly shows that as long as the input is a logic 1, the circuit counts in an infinite loop through states 0, 1, 2, 3, 0, 1, 2, etc. If at any time the input goes to a logic 0, it can be seen that the network maintains the same state for the next clock period.

As another example of the simplicity of a flow diagram, refer to the circuit for a serial gray-code-to-binary converter, shown in Fig. 8-8. This circuit arrangement is called a **serial converter**, because each bit is converted to binary individually, one after another. (This is compared to a **parallel converter**, which converts all of the bits in one clock period.)

It will be recalled from Chapter 2 that the gray-to-binary conversion process consists of comparing each gray code bit to the previous binary bit to determine the state of the new binary

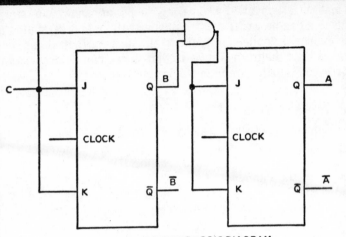

BINARY COUNTER LOGIC DIAGRAM

| PRESENT STATE | NEXT STATE | |
|---|---|---|
| | C = 0 | C 1 |
| A B | A B | A B |
| 00 | 00 | 01 |
| 01 | 01 | 10 |
| 10 | 10 | 11 |
| 11 | 11 | 00 |

BINARY COUNTER TRUTH TABLE

**STATE ASSIGNMENTS**

00 = 0
01 = 1
10 = 2
11 = 3

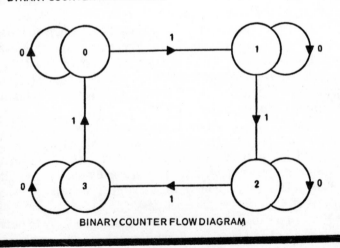

BINARY COUNTER FLOW DIAGRAM

Fig. 8-7. Typical flow diagram.

bit. If the gray code bit is a logic 1, the new binary bit is inverted from the previous binary bit, while if the gray code bit is a logic 0, the new binary bit is the same as the previous binary bit. While the logic diagram and truth table both show this particular implementation, the flow diagram represents the clocked sequential circuit action, regardless of the actual hardware which causes the circuit to function.

## RACE CONDITIONS

Of great concern in sequential circuits is the condition where the input and output variables are in the process of changing states. During this time, the very small delay times through the various gates in a circuit can cause undesired states to occur, resulting in erroneous outputs. For example, an inverter circuit has some finite amount of delay, perhaps a few nanoseconds, which is caused by the storage delay time of the transistor used. Suppose that the input to this inverter changes from a logic 1 to a logic 0 state. There is a short period of time when the inverter will

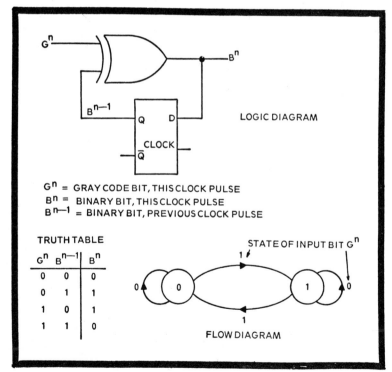

Fig. 8-8. Flow diagram for serial gray-to-binary converter.

have a logic 0 at both its output and its input. Obviously this is a violation of the circuit's proper logic function. During the time when the output does not properly respond to the input, the circuit is called unstable. When the output finally does respond properly, the circuit is said to be stable.

When transistors and resistors are manufactured, there is always some minute variation in their makeup so that by the time complete circuits are built, no two circuits can be guaranteed to have exactly the same delay times. As a result, variables which are intended to change at the same time may not in reality do so.

Consider the case of a standard AND gate, with inputs A and B from some other logic circuits. The desired output F = AB is obtained if A = 1 and B = 1. For explanation purposes, however, assume that the current states of A and B are A = 0, B = 1. Now at some later time, say at the trailing edge of the system clock, the states of A and B both change such that A = 1, B = 0. It is clear that neither the old state nor the new state should produce an output from the AND gate. But due to unequal delays from the various gates, it happens that *A* changes state very slightly before *B* does. Then, for a very brief period, the truth table for the AND gate is satisfied, and A = 1, B = 1, and there is an erroneous logic 1 output. This is an example of a race condition. A race occurs whenever there are two variables attempting to change states at the same time, and due to undetermined delays, the output of a logic network is dependent upon a "race" between the two variables to determine which changes state first. In the case of the AND gate, the race is noncritical because eventually the two variables will settle to their proper states and the logic output will become a 0. If, however, the output of the circuit is unpredictable in that it is entirely dependent on which variable wins the race, then the race is considered a critical one.

An interesting circuit for examining races and stable—nonstable states is the NOR gate SR flip-flop shown in Fig. 8-9. Outputs X and Y correspond to the flip-flop outputs $\bar{Q}$ and Q, respectively. The letters X and Y refer to these outputs after they have had time to respond to the inputs. The small letters $x$ and $y$ refer to the transitional outputs which are cross-coupled back to the circuit inputs. Eventually, if a stable state is to be reached, X must equal $x$ and Y must equal $y$.

A state transition table is made up for this circuit, showing the transitional output states $x$ and $y$ at the left and the resultant new output states X and Y in the individual cells. Note that every possibility of $x$ and $y$ is accounted for. As previously noted, a stable state is achieved any time outputs X and Y are the same as the inputs $x$ and $y$.

Starting at the SR = 11 column of the state transition table, the top cell shows that outputs for X and Y are logic 0s. Further, since $x$ and $y$ are also logic 0s, this is the stable state for the given input. A point of interest here is that, recalling the logic description for a NOR gate SR flip-flop, this is a **not allowed** condition (since the outputs in this case are not complements of one another). Nevertheless, it is the purpose of this description to examine all possibilities; therefore, continuing to the second cell in the column, it is seen that XY = 00 is also contained in this cell. But the old state $xy$ was 01, and it is required that the circuit change to XY = 00. This is accomplished by a change of Y to a logic 0 after a slight time delay.

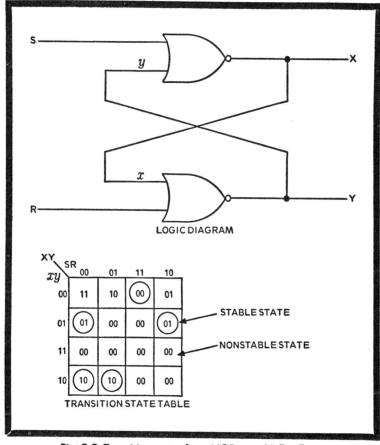

Fig. 8-9. Transition states for a NOR gate SR flip-flop.

If the transition state table is now reconsidered in light of the change, it will be seen that a vertical transition has occurred from the second cell to the first cell. This is the stable state and no further changes will occur. The bottom cell is a similar situation in that it requires only a single-variable change which will cause a vertical transition to the top cell, the stable state.

The third cell in the SR = 11 column, however, calls for a change in the state of $xy = 11$ to $XY = 00$. Here two variables are required to change states simultaneously. This is a race condition which has one of three possible outcomes. If X and Y change at identical times (highly unlikely), then the circuit will immediately reach its stable state and no further changes will occur. But if $x$ changes before $y$, or vice versa, then the circuit will change to the $xy = 01$ or $xy = 10$ state and then go to the $xy = 00$ stable state. It is seen that this is a noncritical race, because no matter which variable changes first, or if they change together, the outcome is the same. That is, the stable state $XY = xy = 00$ is achieved. It should be noted that this particular race condition of $xy = 11$, SR = 11, is itself reached through another race condition for the variables S and R. Another time when this network can achieve the same condition is when power is first applied.

The SR = 01 and SR = 10 columns of the transition state table can be analyzed in a similar manner, with comparable results. In each column, there is one case where both variables are required to change at the same time, and this causes a race condition. As with the SR=11 column, the race is a noncritical race in that the network eventually achieves a known stable state.

The column of particular interest in this transition state table is the SR = 00 column. This column has two stable states. Recalling the logic description of the NOR gate SR flip-flop, it will be remembered that if the flip-flop is in the $XY = 01$ state (set) and the inputs go to SR = 00, the flip-flop remains in the same state.

But trouble arises if the circuit has previously been in the not allowed $XY = 00$ state. If the inputs now switch to SR = 00, one of three results is obtained based upon a race between $x$ and $y$. If $x$ tends to go to a logic 1 first, the circuit will go to the stable state $XY = 10$. If $y$ tends to go to a logic 1 first, the circuit will go to the stable state $XY = 01$. Finally, if $x$ and $y$ should happen to switch at identical times, the circuit will go to the $XY = 11$ state. But this is not a stable state; reapplication of the analysis, using the transition state table, shows that after a slight delay the race will be on again. As before, the circuit can go to either stable state or to the original unstable state. If the latter occurs, the circuit is seen to oscillate between $XY = 00$ and $XY = 11$. Thus, this example is a description of a critical race condition. There is no way of knowing which of the three things the circuit will do. It may oscillate, it may go to the $XY = 01$ state, or it may go to the $XY = 10$ state.

Of course, in operation, this critical race situation is eliminated by making the requirement elsewhere in the logic that $SR = 11$ never be allowed to occur.

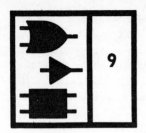

# 9 Typical Logic Networks

There are many different types of logic networks, and every digital system has numerous networks which are unique to that system. There are, however, a few logic networks which are encountered again and again. These networks may vary in minor detail from system to system, but their primary functions are always the same. This chapter introduces a few of the typical logic networks found in almost any digital system. Among the typical logic networks described are registers, counters, adders, and code converters.

## REGISTERS

One of the most important uses for flip-flops in logic networks is to form **registers**. Registers are the means by which digital data is stored for use at appropriate times by other logic networks. The data may be stored for a very short time (1 clock period) to permit its use in some logic operation; or it may be stored for a longer period (several clock periods, where the total elapsed time may be measured in milliseconds) to provide an output after a certain sequence of operations; or it may be stored indefinitely to represent such things as equipment status. Whatever the purpose or time duration of the storage, the same types of registers are used to perform this storage.

Perhaps the simplest type of register which can be configured is the **serial shift** register. This register simply shifts the data bits contained in the individual flip-flops one stage to the right each time a clock pulse occurs. A logic diagram of the basic shift register configuration is shown in Fig. 9-1. This register is made up with JK flip-flops, but there is no reason that D flip-flops or SR flip-flops could not perform the identical functions. To describe the operation of this circuit, consider the timing diagram shown directly below the logic diagram. A serial data stream is presented at the shift register data input. The data stream in this example consists of the bits 1101, which occur sequentially, the most significant bit first. Assuming that the register is initally in the 0000 state, a logic 1 is read into flip-flop A by the first clock pulse, causing the state of the register to become 0001. At the second clock pulse, the data is shifted right one bit, so that the contents of flip-flop A are now contained in flip-flop B. At the same time, the next serial data bit is read into flip-flop A. Thus, after clock pulse two, the state of the register is 0011.

The third clock pulse produces a similar right shift, changing the register state to 0110. The fourth clock pulse reads in the final data bit, while again shifting data in the other flip-flops one place to the right, and the state of the register becomes 1101. At this

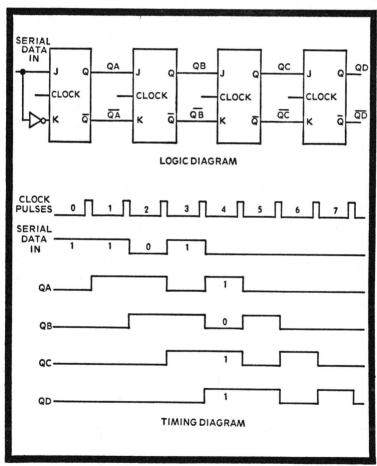

Fig. 9-1. Basic shift register configuration.

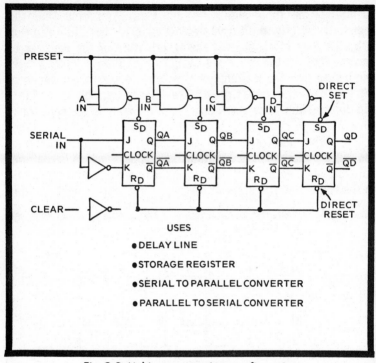

Fig. 9-2. Multipurpose register configuration.

point, the serial data is now fully contained in the register and a number of options are available:

a. The individual flip-flop outputs may be read out in parallel during the fourth clock period. In this manner, the data can be transferred to other logic circuits.
b. The clocks to the register can be stopped. Here, the register is used as a temporary storage register, storing the data until needed.
c. The output of flip-flop D can be used as a delayed serial data stream. In this case, the register acts as a four-clock-period delay line.

A more general-purpose register configuration is shown in Fig. 9-2. This register features a direct set and reset capability which is frequently found in integrated circuit flip-flops and registers. Basically, the direct set and reset inputs override all other inputs to enter data into the flip-flops. Thus, for example, no matter what actions are occurring at the JK inputs, if a pulse is applied to the clear input, the register will assume the 0000 state.

Similarly, once having been cleared, any desired data bits may be parallel entered into the register through the A, B, C, and D inputs, if a logic 1 is also present at the preset input. There is no requirement that a clock pulse be present. Thus, the direct set and reset inputs allow parallel data entry into the register, with a resultant increased flexibility for various uses. Consider, for example, some typical uses of this register:

## Delay Line

If operated as a serial shift register, the register acts as a delay line, providing a delay equal to the number of flip-flops in the register. One frequency application of a digital delay line is to align several different sets of digital data so that their bit positions agree. After alignment, arithmetic or logic operations may be performed on the data.

## Storage Register

Data can be entered in serial form or through the parallel inputs for storage. As long as the flip-flops receive no clock inputs after the data is in the register, the flip-flops will store the data for as long as desired. Alternatively, the output of flip-flop D can be connected back to the register input to provide a recirculating register that shifts the data in a continuous loop. Care must be taken to keep track of the number of shift clocks which have been received, so that the proper location of data can be determined.

## Serial-to-Parallel Converter

Serial data is shifted into the register until it is properly aligned, then read out in parallel format. A common reason for converting from serial to parallel data format is so that logic operations can be performed in a shorter period of time. Since each bit of serial data requires one clock period to become available, it can be seen that a simple function like adding two 10-bit numbers requires 10 clock periods. In parallel format, this operation requires only one clock period.

## Parallel-to-Serial Converter

By using the direct set and reset inputs to this circuit, parallel data can be entered into the register. If clocks are now applied to the register, the output of flip-flop D is a serial data stream representing the parallel input data. One reason for converting to serial data format is the transmission of digital data. In parallel format, a 10-bit number requires 10 line driver circuits and 10 wires in a cable. In serial format, only one driver and one wire is required.

## Counters

In addition to registers, almost any digital device will have one or more counters. Counters, like registers, are also made up of flip-flops and associated gating. In general, the maximum number of states which a counter can have is $2^n$, where n is the number of flip-flops in the counter. For example, the counter shown in Fig. 8-7 has two flip-flops, to obtain $2^2 = 4$ states. A three-stage counter could have a maximum of $2^3 = 8$ states. The actual number of counts defined by a particular design is called its *modulo* count. For example, a *modulo seven* counter has 7 counts in its count sequence.

There are two basic types of counter construction: ripple and synchronous. A ripple counter is the simpler to build and uses the output from each preceding stage as the clock for the next stage. However, if there are quite a few stages to the counter, an excessive amount of delay time may be encountered between the time the first stage changes states and the time when the last stage performs its transition.

To overcome this problem, synchronous counters are used when speed is an important factor. A synchronous count is obtained by using additional logic gates to insure that all stages of the counter change states at the same time

## Ripple Counter

An example of a typical ripple counter is shown in Fig. 9-3. This counter is *modulo eight* because it has 8 counts and it is binary because it counts in a pure binary sequence. It is a ripple counter because the clock for each stage is the output from the previous stage.

There are several points which should be kept in mind when analyzing this circuit: First, it should be remembered that flip-flops change states at either the leading or the trailing edge of the clock pulse. The flip-flops shown here change states on the negative-going clock pulse edge. Secondly, a key to this circuit's operation is the ability of a JK flip-flop to toggle if its J and K inputs are both a logic 1. Since positive-true logic is being used throughout this book, $+V$ represents the logic 1 condition. Hence, it is seen that each stage toggles, and the rate of toggling is always at half the clock frequency. Thus, flip-flop A divides by two, flip-flop B divides by four, and flip-flop C divides by eight.

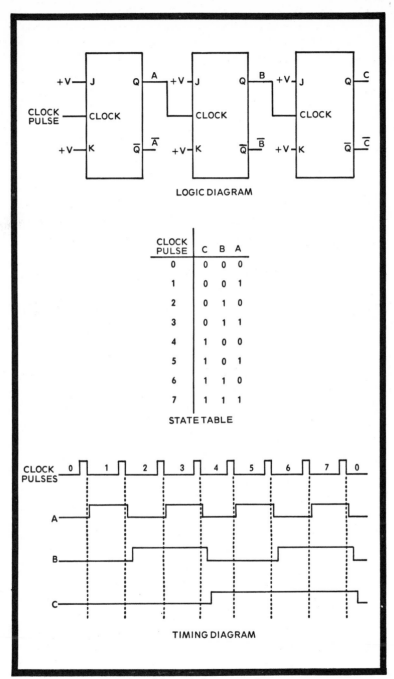

Fig. 9-3. Modulo-eight binary ripple counter.

From the timing diagram, it is seen that each successive division occurs on the trailing edge of the clock pulse. Note also that all the counts found in the state table can be read from the timing diagram, if desired. The state table merely provides a more convenient and compact notation for the information; and usually, the minute details of the timing diagram need not be considered to understand the operation of the counter.

## Synchronous Counter

A synchronous counter is shown in Fig. 9-4. This counter is *modulo sixteen* because it has 16 counts and is binary because it counts in a binary sequence. The counter is synchronous because all four stages change states at the same time, at the trailing edge of the clock pulse. Thus, there is no ripple delay time as was the case in the ripple counter.

The operation of this counter is quite different, in that instead of toggling the flip-flop on a trailing edge of some signal generated by a previous flip-flop, a desired state is detected and the flip-flop is caused to toggle on the next clock pulse after the detected state occurs. Since the previous stage also has its transitions on the trailing edge of the clock pulse, the two stages change states at the same time, or *synchronously*.

Examining the state table, each clock pulse is seen to cause flip-flop A to toggle. Flip-flop B toggles one clock period after flip-flop A has been set to a logic 1. Similarly, flip-flop C is caused to toggle one clock pulse after flip-flops A and B are both set to the logic 1 level. Flip-flop D operates in the same way, requiring that a change of state take place one clock pulse after detecting that flip-flops A, B, and C are all set to the logic 1 state. Each of these requirements can be readily seen by noting the count sequence which the state table must follow if a binary count is to be obtained.

Interestingly, even though the techniques for binary and ripple counters are quite different, their state tables are identical. As previously mentioned, the main difference between the two is their speed of operation.

## BCD (Decade) Counter

The counters previously described all used the maximum number of counts available, that is, $2^n$ states. However, counters can readily be made to count any desired number of counts by providing the proper gating to skip some counts which would

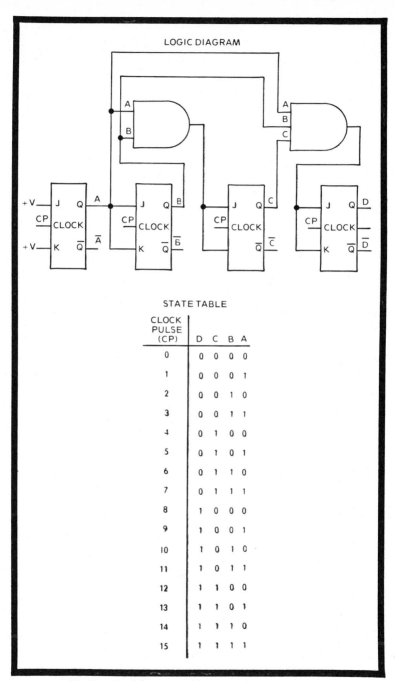

Fig. 9-4. Modulo-sixteen binary synchronous counter.

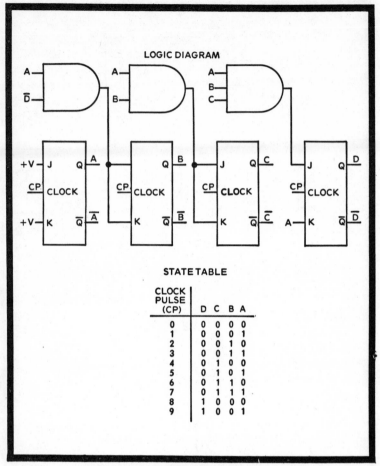

Fig. 9-5. 8421 BCD synchronous counter.

normally occur. One such counter which is commonly used is the BCD or decade counter. A decade counter is a *modulo ten* counter (because it has 10 counts).

A typical decade counter which counts in 8421 BCD is shown in Fig. 9-5. This counter is very similar to the synchronous *modulo sixteen* counter previously described, except that the logic gating has been modified in such a way that the last six counts are skipped.

To see the skip action, consider the natural binary sequence presented by the *modulo sixteen* counter. As the counter reaches the ninth count (1001), the next state is the tenth count (1010); but instead, it is desired that the next count in sequence be zero (0000). The states of flip-flops A and C will be correct and need not be changed. Flip-flop B, however, will toggle to the 1 state unless this action is inhibited. Thus, the toggle term for flip-flop B is modified from $J = A$, $K = A$ to $J = A\bar{D}$, $K = A\bar{D}$. here, $\bar{D}$ stops flip-flop B from toggling after the count reaches nine.

Flip-flop D presents a different problem in that it is already set and must be reset after the ninth count. Since the only time that flip-flop D needs to be set is after the seventh count, and the only time it needs to be reset is two counts later, separate set and reset terms are called for. The set term is count seven, the same logic gating which was previously used. The reset, however, can occur each time flip-flop A goes set and, in particular, it is required at the clock pulse immediately following the ninth count. By making these simple modifications, the desired 10 counts are obtained in the 8421 code and the remaining 6 counts are skipped.

### Johnson Counter

A very simple counter which uses no logic gates is the Johnson counter. A Johnson counter is basically a shift register which has a very unique feedback to its serial input. The feedback is configured such that whatever the state of the output stage, the complement of that state is applied to the serial input at the next clock pulse.

A *modulo ten* Johnson counter is shown in Fig. 9-6. Note from the state table that logic 1 levels are shifted into the register for five counts, then logic 0 inputs are shifted for the next five counts, etc. An important feature of the Johnson counter is that the cycle length is 2n, instead of $2^n$. Hence, for the example shown there are five stages, so that $n = 5$, and the resultant cycle length is $2n = 10$. Because there are 2n counts and $2^n$ possible states which $n$ flip-flops can assume, the counter can be in any one of these 2n states when power is applied.

If the state which occurs is not one of those in the normal count sequence, the counter will not count properly and it may never correct itself to the proper sequence. One method for insuring that the counter will operate properly is with a **clear** input. Upon application of power, a logic 1 is momentarily applied to the clear line. This sets the register to all zeroes; as seen from the state table, this is one of the allowed counts. Upon removal of the logic 1 from the clear input, the counter will then count correctly.

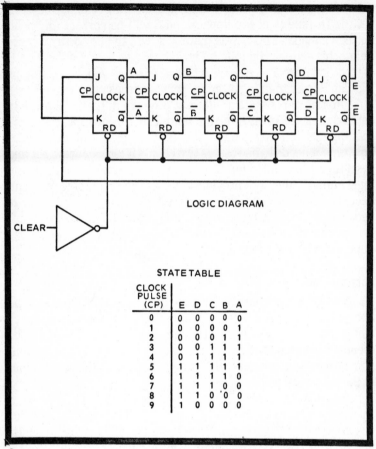

**Fig. 9-6. Johnson decade counter.**

## Ring Counter

A ring counter is another version of a shift register configuration which uses significantly less than the maximum number of states which can occur. The counter is simply a recirculating shift register into which a single 1 has been entered. Thus, for $n$ flip-flops, the counter has $n$ states. A typical ring counter is shown in Fig. 9-7. From the state table, it is apparent that each count has a single 1 and four 0s. The 1 shifts from stage to stage in an endless loop.

As with the Johnson counter, a ring counter must be initialized before it will operate properly. A method which can be used in the counter shown consists of presenting a clear pulse followed by a preset pulse to the input lines. When the clear pulse

**124**

occurs, the counter will go to the 00000 state. The preset inputs are wired so that when the preset pulse occurs, the counter assumes the 00001 state. From this point on, the counter operates in its normal sequence for each clock pulse received. If at any time it is desired to restart the counter at its initial count, the above procedure must be repeated.

## DECODERS

Many applications for the counters just described use the flip-flop outputs directly to perform some type of control function or to form a binary or BCD number. However, often a requirement exists to know when a particular count of the counter occurs and to use this count to start or terminate a particular sequence of events. The method by which the occurrence of a specific count is known is by decoding the desired counter state. Each decode is simply an AND function with its inputs equal to the state being decoded.

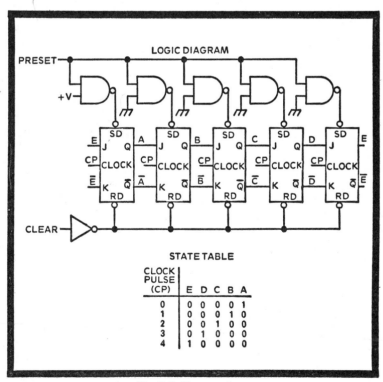

Fig. 9-7. Ring counter.

## Single Decode Gate

An example of a typical decode gate is shown in Fig. 9-8. This particular gate represents a decode of count nine of the *modulo sixteen* binary counter shown in Fig. 9-4. Since a logic 1 is desired for only count nine, the Karnaugh map has only one cell with a logic 1. Also, assuming that the *modulo sixteen* counter is running continuously, the timing diagram shows the resultant decode gate output for count nine. Obviously, the logic equation can be written very simply from either the truth table or from the Karnaugh map. It is also evident that any other count can be decoded in a similar manner.

## Don't-Care Conditions

An important criterion when determining the requirement for a decode gate is whether or not unused states of the counter exist. Consider, for example, the truth table and associated Karnaugh map for an 8421 BCD count sequence, as shown in Fig. 9-9. It is apparent that only the first 10 out of 16 possible states are used. The remaining states can therefore be considered unused states. If indeed these states never occur, then surely it does not matter if a decode gate is formed such that it will produce an output for one or more of the nonexistent states. Thus, the outputs for the six unused states in this example are called *don't care* outputs, and they are designated by Xs in the truth table and Karnaugh map. The beauty of *don't care* conditions is that when loops are drawn in the Karnaugh map, as many Xs as desired can be included in the loop. It has already been shown that the larger the loop, the fewer gates and gate inputs required to implement the function. Including selected *don't care* conditions in the Karnaugh map loop is equivalent to defining that if the unused states should occur (which they won't), then a logic 1 output will be obtained from the decode gate.

Referring to the Karnaugh map, each *f* in the map represents a desired output of logic 1 for that count alone. Thus, f9 means that a logic 1 output is desired when count nine occurs. Note that for counts 0 and 1, a normal four-input decode is required. However, counts 2, 3, 4, 5, 6, and 7 all include Xs to form a two-variable loop; thus, gates can be formed to decode these counts, which each require one less input.

Finally, it is observed that for counts 8 and 9, a loop of four cells can be made. Thus, these two decodes require only two inputs. From this, we see that it is important to recognize *don't care* conditions and to plot these conditions in the Karnaugh map whenever decode functions are analyzed.

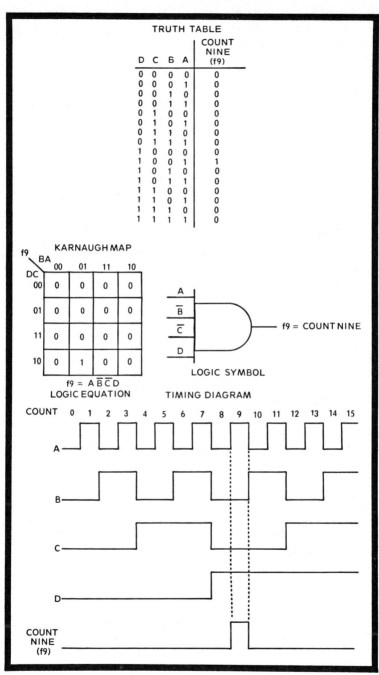

Fig. 9-8. Basic decode gate.

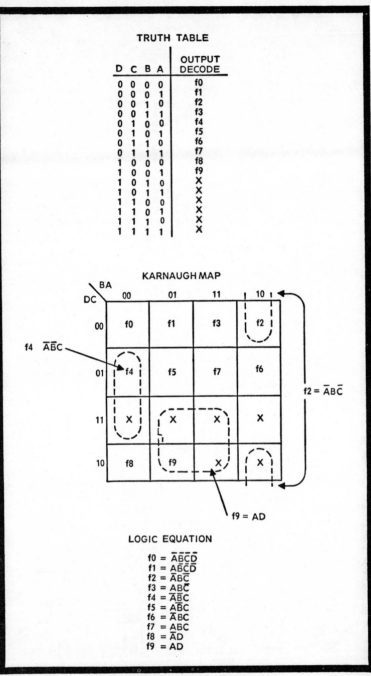

**TRUTH TABLE**

| D | C | B | A | OUTPUT DECODE |
|---|---|---|---|---|
| 0 | 0 | 0 | 0 | f0 |
| 0 | 0 | 0 | 1 | f1 |
| 0 | 0 | 1 | 0 | f2 |
| 0 | 0 | 1 | 1 | f3 |
| 0 | 1 | 0 | 0 | f4 |
| 0 | 1 | 0 | 1 | f5 |
| 0 | 1 | 1 | 0 | f6 |
| 0 | 1 | 1 | 1 | f7 |
| 1 | 0 | 0 | 0 | f8 |
| 1 | 0 | 0 | 1 | f9 |
| 1 | 0 | 1 | 0 | X |
| 1 | 0 | 1 | 1 | X |
| 1 | 1 | 0 | 0 | X |
| 1 | 1 | 0 | 1 | X |
| 1 | 1 | 1 | 0 | X |
| 1 | 1 | 1 | 1 | X |

**KARNAUGH MAP**

f4  $\overline{A}\overline{B}C$

$f2 = \overline{A}B\overline{C}$

$f9 = AD$

**LOGIC EQUATION**

$$f0 = \overline{A}\overline{B}\overline{C}\overline{D}$$
$$f1 = A\overline{B}\overline{C}\overline{D}$$
$$f2 = \overline{A}B\overline{C}$$
$$f3 = AB\overline{C}$$
$$f4 = \overline{A}\overline{B}C$$
$$f5 = A\overline{B}C$$
$$f6 = \overline{A}BC$$
$$f7 = ABC$$
$$f8 = \overline{A}D$$
$$f9 = AD$$

Fig. 9-9. Decodes for 8421 BCD counter.

### Decodes for Johnson and Ring Counters

Both the Johnson counter and the ring counter typically have many more states than their modulo counts. Since it has just been shown that unused states result in *don't care* conditions, it might be expected that the very large number of *don't cares* would simplify decoding considerably. This is exactly the case. Decodes for these two counter configurations are shown in Fig. 9-10.

For the Johnson counter it can be seen without even looking at a Karnaugh map that no matter what the Johnson cycle length, it is always possible to decode a particular count with a two-input AND gate. The two inputs to the gate are always such that the decode follows the $1 - 0$ or $0 - 1$ changeover. The

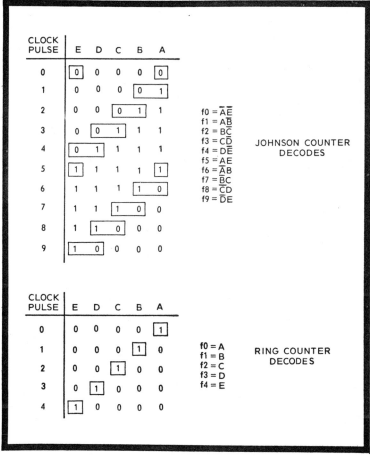

Fig. 9-10. Decoding the Johnson and ring counters.

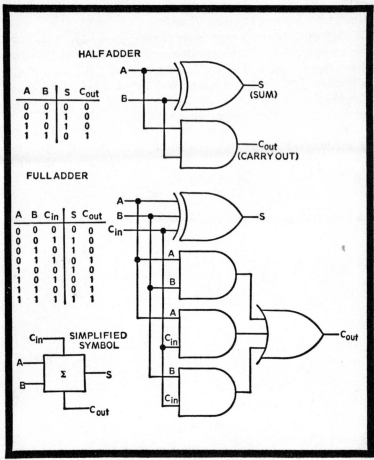

**Fig. 9-11. Basic adder configurations.**

exceptions to this are the all-0 state and the all-1 state. Here, the two end stages of the counter always form the decode.

The decode for a ring counter is even simpler. In fact, it isn't really a decode at all: Since there is always a single 1 in the counter, the stage which has the 1 in it represents the state of the counter. Thus, no gating is required to decode a ring counter. All of the above description can be derived from looking at Karnaugh maps and drawing loops, but in simple cases such as this, intuitive common sense is the simpler approach.

## ADDERS

Another common logic network is the *adder* circuit. There are usually two forms of the adder in common applications. The first is a *half adder*, which adds two binary bits and generates a sum and a carry output. The second form is the *full adder*, which adds two binary bits plus a carry input to produce the sum and carry outputs. The two circuits are shown in Fig. 9-11.

Consider first the *half adder* circuit. The truth table shows that this is the standard binary addition first identified in Chapter 2. The sum is simply an exclusive-OR function of inputs A and B and a carry is generated if two 1s are added.

But the situation is not all that simple if more than two bits are to be added. In this case, there is a possibility of a carry input to the addition from a previous bit. Hence, binary addition becomes a three-variable problem, with outputs as shown in the *full adder* truth table. It is seen that the sum output is the exclusive-OR of all three inputs, while a carry bit is generated any time more than one of the inputs is a logic 1.

Instead of drawing all of the logic each time an adder is used, the simplified symbol shown will be employed. If the symbol has a carry input, then a full adder is implied; otherwise, a *half adder* can be used.

### Serial Adder

A block diagram for a serial adder is shown in Fig. 9-12. Some binary number, A4A3A2A1, is stored in register A. A second number, B4B3B2B1, is stored in register B. The process of addition begins by resetting the D flip-flop so that the carry bit will be a logic 0 to start with. The two registers now shift out their contents, both at the same time, while register C is simultaneously shifting in sum bits as they are generated. For example, suppose that the two numbers to be added are as shown in the diagram. At the first clock pulse, the inputs A and B are both logic 1 and there is no carry in. Therefore, the first sum bit, a logic 0, is shifted into register C and a logic 1 carry bit is stored in the D flip-flop. At the second clock pulse, the carry bit will be present at the carry input and inputs A and B will be 0 and 1, respectively.

According to the truth table rules, the sum bit will again be a logic 0 and another logic 1 carry bit will be generated. Following this procedure through the fourth clock pulse results in a logic 1 carry bit being left in the D flip-flop and no further data bits left in the A and B registers. But the carry bit is a perfectly valid part of the sum; therefore, one last clock pulse must be used to complete the serial addition. This example demonstrates that, when adding two binary numbers, the sum register must always

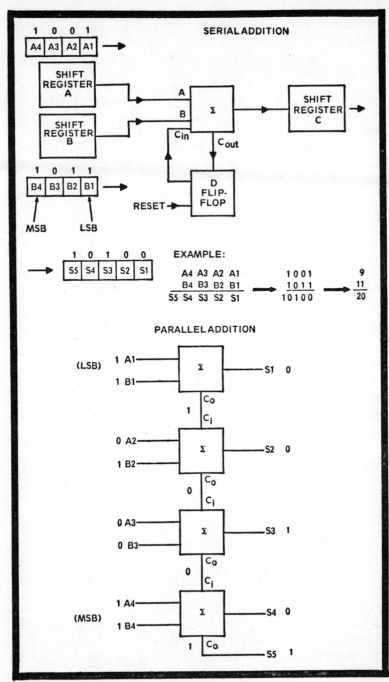

Fig. 9-12. Serial and parallel adders.

be able to accommodate one more bit than was in the input registers.

Registers A, B, and C are normally existing registers, used for multiple functions in the system, and the only extra logic required, in addition of this type, is the full adder circuit and a single flip-flop.

The main advantage to a serial adder is its simplicity and the minimum amount of logic required. Its primary disadvantage is that it takes many clock periods to complete one addition; so if there are many such additions to be performed, the time required may become excessive. A parallel adder can perform this same operation in less than one clock time.

### Parallel Adder

The parallel adder shown in Fig. 9-12 uses the same full adder circuit as the serial adder, but adds all bits at the same time. As can be seen, this configuration requires one complete full adder circuit for each bit position. If large numbers are being added, this represents a considerable amount of logic. Further, the parallel adder still requires registers A, B, and C in which to store its results.

The big advantage to parallel adders is that the entire operation can be performed in nanoseconds (thousandths of microsecond) as compared to microseconds (millionths of a second) for the serial adder.

The same example as was used for the serial adder can be used to analyze the parallel adder circuit. The two inputs A1 and B1, as before, are each at logic 1, resulting in S1 = 0 and a carry of 1. At the same time, inputs of A2B2 = 01 are also present at the second adder. This generates a sum S2 = 0 and again a logic 1 carry is propagated to the next stage. The action described is repeated for each parallel stage until the final carry is obtained. The maximum speed of operation for this circuit is determined by the length of time it takes a carry to propagate from the first adder (least significant bit) to the carry out of the last adder (most significant bit).

### CODE CONVERTERS

Throughout this book, quite a few different binary codes have been described, each having unique advantages. Because any particular digital system may use one or more of these codes, a frequently encountered requirement is for a logic network that can convert from one code format to another. A circuit which performs this function is called a *code converter*.

Several code converters have already been described. For example, the collection of 10 decodes for the 8421 BCD counter, shown in Fig. 9-9, is often referred to as a BCD-to-decimal code converter. The inputs to the code converter are the four BCD bits, and its outputs are the ten decimal digits.

Another code converter which was previously described is the serial gray-to-binary code converter shown in Fig. 6-9. Many other code converters are possible, several of which are described in the following paragraphs.

### Decimal-to-BCD Converter

One reason for converting a BCD number to decimal might be to light some decimal display indicating which of 10 possible actions had occurred in a circuit. The decimal readout can then be interpreted by someone not familiar with binary number codes. Similarly, occasion arises to enter decimal data via switches into the digital circuits. Ten bits of data are very wasteful when the same information can be represented by four bits in a BCD code. For this, a decimal-to-BCD code converter similar to that shown in Fig. 9-13 is used.

This circuit is extremely simple. Each bit in the BCD code is set to a logic 1 by several decimal digits. Thus, to encode the decimal number into a BCD number, all of the decimal digits which produce a logic 1 for a particular bit position are ORed together. The circuit shown is a code converter for 8421 BCD; however, any BCD code can be formed in an identical manner.

### Gray-to-Binary Converter

Two different methods for performing a conversion from gray code to binary are shown in Fig. 9-14. The serial converter has already been described in Chapter 8. Recalling, a serial data stream is presented at the gray code input, most significant bit first. The binary output is an exclusive-OR of the current gray-code bit and the previous binary bit. This is another way of stating that if the current gray-code bit is a logic 0, the current binary bit will be the same as the previous binary bit. If the current gray-code bit is a logic 1, the current binary bit will be the complement of the previous binary bit.

With each clock pulse, a new gray-code bit is presented at the input and a new binary bit is generated at the output. As has been previously mentioned, one disadvantage to serial operations is the large number of clock periods required to perform a given logic function. This is also true for the serial gray code converter.

An alternate approach to gray code conversion is a parallel code converter. In this circuit, all bits are present at the input, and in a matter of nanoseconds, all outputs are also available. The disadvantage to a parallel converter is that many logic gates are required to perform a function which previously required only one gate.

## BCD-to-Binary Converter

There are many ways of performing a BCD-to-binary conversion. Two commonly used methods are shown in Fig. 9-15.

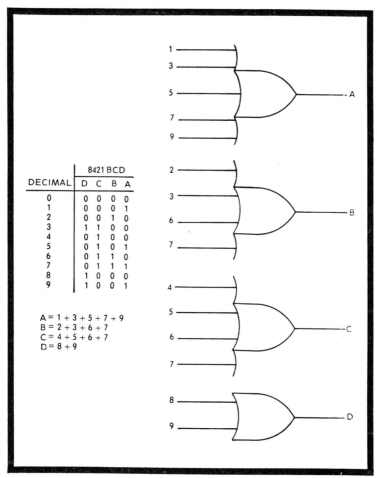

| DECIMAL | 8421 BCD | | | |
|---------|---|---|---|---|
| | D | C | B | A |
| 0 | 0 | 0 | 0 | 0 |
| 1 | 0 | 0 | 0 | 1 |
| 2 | 0 | 0 | 1 | 0 |
| 3 | 1 | 1 | 0 | 0 |
| 4 | 0 | 1 | 0 | 0 |
| 5 | 0 | 1 | 0 | 1 |
| 6 | 0 | 1 | 1 | 0 |
| 7 | 0 | 1 | 1 | 1 |
| 8 | 1 | 0 | 0 | 0 |
| 9 | 1 | 0 | 0 | 1 |

A = 1 + 3 + 5 + 7 + 9
B = 2 + 3 + 6 + 7
C = 4 + 5 + 6 + 7
D = 8 + 9

Fig. 9-13. Decimal-to-BCD code converter.

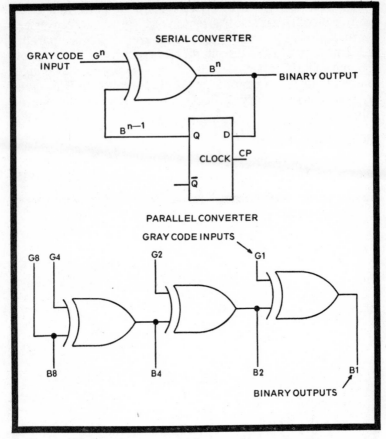

Fig. 9.14. Two versions of gray-to-binary converter.

The serial converter method is the simplest of the two but, as with other serial operations, it is also the slowest. A BCD number to be converted is preset into a BCD *down counter*, while at the same time, a binary *up counter* is reset to the all-0 state. At the next clock pulse, the clock control flip-flop is set and clock pulses are allowed to pass through the AND gate to the two counters. At each following clock pulse, the BCD counter will count down in a reverse binary sequence towards zero.

Meanwhile, the binary counter will count up (away from zero) at each clock pulse. The number of clock pulses for the BCD counter to reach count zero is equal to the BCD count which was preset into the counter. However, since the binary counter is also counting the same number of clock pulses, the correct binary count will continually be maintained by the binary counter. When

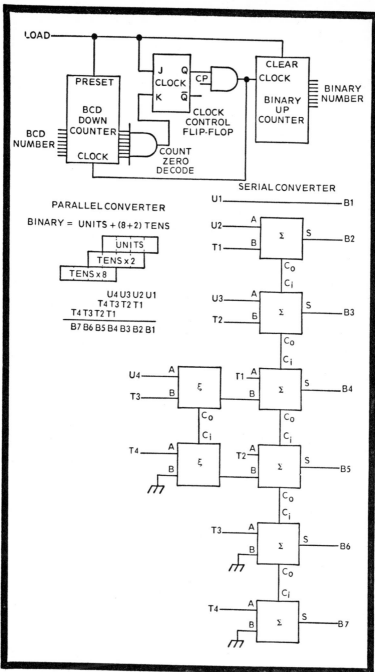

Fig. 9-15. Two versions of BCD to binary converter.

the BCD counter reaches count zero, the count-zero decode gate resets the clock control flip-flop, thus discontinuing the count sequences of both counters. At this time, the numerical value contained in the binary counter can be read and it will be equivalent to the original BCD number.

A more complex method, which can perform the conversion in less than one clock period, is through the use of parallel adders. Any BCD number can be thought of as consisting of a *units* decade plus a *tens* decade, multiplied by a sum of powers of two which is equal to 10. As an arithmetic expression, binary = units + tens $(8 + 2)$. If the units and tens are summed in an appropriate manner, the result will be a binary number consisting only of power of two. As an example, consider the conversion of the BCD representation for 69 to a comparable binary representation. The BCD representation is $T4T3T2T1$ $U4U3U2U1 = 0110\ 1001$. Performing the summations described produces the results shown below:

| | | | U4 | U3 | U2 | U1 |
|---|---|---|---|---|---|---|
| | | T4 | T3 | T2 | T1 | |
| T4 | T3 | T2 | T1 | | | |
| B7 | B6 | B5 | B4 | B3 | B2 | B1 |

→

| | | | 1 | 0 | 0 | 1 |
|---|---|---|---|---|---|---|
| | | 0 | 1 | 1 | 0 | |
| 0 | 1 | 1 | 0 | | | |
| 1 | 0 | 0 | 0 | 1 | 0 | 1 |

The arithmetic just shown is the same arithmetic performed by the parallel converter circuit. A similar converter could be made up for a three-decade BCD number by recognizing that binary = units + tens $(8 + 2)$ + hundreds $(64 + 32 + 4)$. It can be seen, however, that the more decades in the BCD number, the larger the number of adders required. As a result, the technique shown is normally not useful beyond three BCD decades.

# Complex Functions In Intregrated Circuits

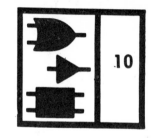

Prior to the advent of integrated circuits, digital logic was built from individual transistors and resistors. Thus, these were the components with which the designer and maintenance man were familiar. Later, many digital systems were built using integrated circuits, manufactured such that several gates or flip-flops were contained in one package. In these systems, the concern was with the individual logic functions.

As integrated circuit technology has grown, greater packaging densities have been achieved, and with these higher densities, it has become feasible to include complex functions in a single package. The packaging density is considered *medium-scale integration* (MSI) if between 25 and 100 gates are contained in a single package. *Large-scale integration* (LSI) encompasses packaging densities where hundreds of gates are included in a single package.

It is the purpose of this chapter to acquaint the reader with some of the complex functions currently included in modern digital systems. It should be remembered that there are many different logic families and many manufacturers of integrated circuits. Thus, no effort is made here to precisely represent any specific implementation of a complex function, but rather to show the various types of functions likely to be encountered.

## Counters

Counters are among the most common complex logic functions available. Typical of the configurations encountered are binary and BCD counters operating either synchronously or as ripple counters. These counters are most frequently found in the DTL, TTL, ECL, and MOS logic families. An example of the advantages which can be obtained with MSI counters is shown by a standard up—down dinary counter.

A truth table for a four-stage up—down counter is given in Table 10-1. From this truth table, J and K terms for individual

**Table 10-1. Truth Table for Four Stage Up / Down Binary Counter.**

| PRESENT STATE | NEXT STAGE | |
|---|---|---|
| | COUNT UP | COUNT DOWN |
| 0000 | 0001 | 1111 |
| 0001 | 0010 | 0000 |
| 0010 | 0011 | 0001 |
| 0011 | 0100 | 0010 |
| 0100 | 0101 | 0011 |
| 0101 | 0110 | 0100 |
| 0110 | 0111 | 0101 |
| 0111 | 1000 | 0110 |
| 1000 | 1001 | 0111 |
| 1001 | 1010 | 1000 |
| 1010 | 1011 | 1001 |
| 1011 | 1100 | 1010 |
| 1100 | 1101 | 1011 |
| 1101 | 1110 | 1100 |
| 1110 | 1111 | 1101 |
| 1111 | 0000 | 1110 |

flip-flops could be derived and a counter built out of gate and flip-flop packages. Such a circuit would require many packages to implement; and even then, if a failure occurred, detailed troubleshooting procedures would be required to isolate the faulty component. As a complex function, the entire circuit is contained in one package with a single line controlling whether the counter counts up or counts down. No external logic is necessary. Furthermore, if a failure should occur, it is merely necessary to determine that the entire counter is not functioning properly and to replace the counter package.

Also, because more functions can be included in the package, a very common feature is the inclusion of preset inputs for each individual flip-flop. With these inputs, it is a simple matter to program a given counter to count any desired modulo. For example, two different *modulo twelve* counters are shown in Fig. 10-1. Both are made up from *modulo sixteen* counter packages, with four counts skipped. If a binary count is desired, the recommended technique is to decode the end count of that particular modulo sequence, then preset the counter to the all-0 state. This simple change requires one gate.

A second method is by using the terminal count output which is often provided. In the example shown, the terminal count is defined as count fifteen. By using the terminal count to preset the counter, any number of the earlier states can be skipped. Here,

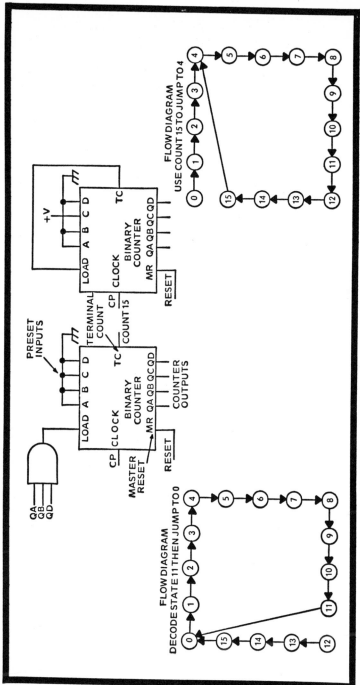

Fig. 10-1. Two different modulo-twelve counters.

states 0, 1, 2, and 3 are skipped. The disadvantage to this arrangement is that the count does not start at 0000 and progress in a normal binary fashion. Thus, any decodes taken from the counter must take this into account. Say, for example, that the second count of the sequence shown is to be decoded. State five (0101) would be the decode necessary to obtain the proper output.

## STATIC SHIFT REGISTERS

The register configurations described in Chapter 9 can all be referred to as static registers. That is, they are all made up from flip-flops which can hold data indefinitely. Similarly, shift registers are static shift registers if clocks can be stopped without losing any data. Static shift registers can be restarted and the stored data can then be shifted out, or new data entered. All of the static register types are made in MSI configurations. The registers are usually 4, 5, 8, or 16 bits long and can be of the shift register type or the storage register type.

Almost any combination of serial—parallel entry or serial—parallel readout is available. Usually, the longer registers are of the serial-in—serial-out type. This is simply due to the fact that too many input and output leads would be required to make all of the intermediate stage inputs and outputs available.

## DYNAMIC SHIFT REGISTERS

If the clocks to a shift register must run continuously, the register is considered a dynamic shift register. Such is the case for the MOS dynamic storage described in Chapter 5. Recalling, the method of storage used is the gate capacitance of a MOS inverter stage, using a MOS gate to transfer the stored charge from one stage to another. If the clock to this type of shift register were stopped, the capacitive storage would eventually decay and data would be lost. Very large amounts of delay can be obtained through the use of MOS dynamic shift registers. The number of bits of delay runs from about 25 to more than 1000 bits. This is obviously large-scale integration.

A two-stage dynamic shift register is shown in Fig. 10-2. A two-phase clock is used so that data can be transferred a half-stage at a time. This permits data to be copied into one half-stage without destroying the data contained in the other half-stage. The phase one clock can be considered an *intermediate transfer* clock, while the phase two clock can be considered the *read in—read out* clock.

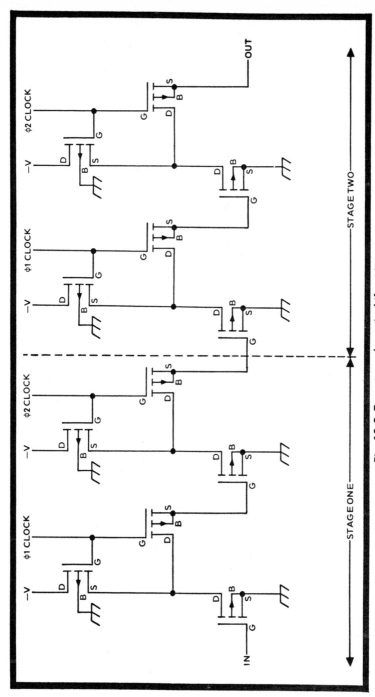

Fig. 10-2. Two-stage dynamic shift register.

143

Assume that an input data bit is present at the gate to the first inverter. When the phase one clock pulse occurs, the second inverter gate capacitance begins to charge. As soon as the phase one clock pulse goes away, the charge is stored in the second inverter of the first stage. Thus, the phase one clock has temporarily transferred the data within the first stage. When the phase two clock pulse occurs, the second-stage inverter gate capacitance charges.

Now, when the phase two clock is removed, the data is stored in the second stage. Hence, the phase two clock has read the data out of the first stage and into the second stage.

## MULTIPLEXERS

A digital multiplexer selects one signal from among several inputs and applies this signal to the output. The selection of the appropriate signal is controlled by an *input select* code. One typical use for multiplexers is the sequential (serial) application of a number of signals onto a single line. If this process is strictly a function of time (that is, if each line is sequentially selected), the process is referred to as *time division* multiplexing. Functionally, time division multiplexing might be considered another method of parallel-to-serial conversion.

Another use for digital multiplexers is data routing. Suppose, for example, that one signal is desired in a particular operating mode, while some other signal is desired if the mode is changed. The multiplexer provides the means for making this data selection.

An example of an eight-input multiplexer is shown in Fig. 10-3. The entire logic array is simply a group of AND gates which decode the select lines and apply the appropriate input signal to an OR gate at the output. A number of inverters are included to obtain the complements of the select lines and to provide buffering. Of course, the reason for inclusion in this chapter is that the multiplexer is available as a complex function in the MSI class of integrated circuits. As a complex function, it is observed that many interconnections and individual gates are combined into a single package, thus simplifying the design and maintenance functions.

## MEMORY ELEMENTS

Semiconductor memory arrays are of two basic types, *serial data access* and *parallel data access*. Static and dynamic shift registers are examples of serial memory organizations. Storage

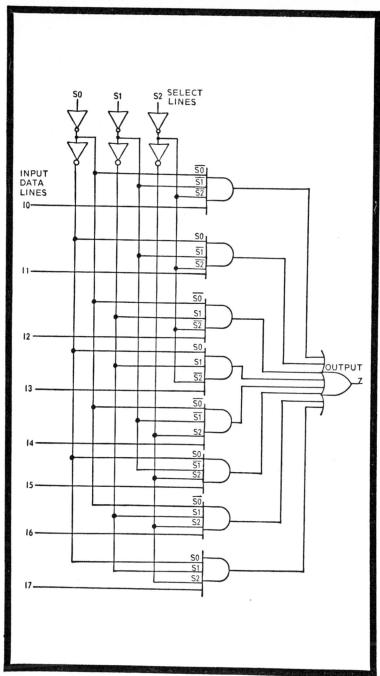

Fig. 10-3. Eight-input multiplexer.

145

registers and decoding matrixes are examples of parallel memory. Often, memory requirements become quite large, such that large-scale integration techniques are justified. Using modern technology, relatively sophisticated data entry and retrieval schemes can be included in a single integrated-circuit package. The descriptions that follow describe memory arrays with up to several-thousand-bit capacities, which are typically found in a single package. However, larger memories can always be formed by interconnecting a number of such packages.

## Read-Only Memories

A *read-only memory* (ROM) is simply a very large array of gates wired such that upon application of an address consisting of several bits, a fixed set of outputs is obtained. The array is a memory, because it "stores" a known set of outputs for a given input condition. The memory is known as *read-only* because no new data can be written into it.

The memory is wired in a fixed manner. Read-only memories are very useful as lookup tables to perform code conversions or to look up mathematical functions. Typical mathematical functions include sine, cosine, or logarithmetic lookups or even binary multiplication tables.

To demonstrate the basic simplicity of the read-only memory concept, a simplified schematic of a diode memory, wired to perform decimal-to-binary conversion, is shown in Fig. 10-4. This circuit has memory, because stored in the logic array (the diodes plus wiring) is a binary number for every decimal number that can be applied at the input. The inputs are decoded in a standard manner, forming four bits representing the powers of two which correspond to the decimal number applied.

For each output line, a diode connection is made where a logic 1 output is desired and no connection is made where a logic 0 output is desired. As can be seen, this read-only memory represents a logic circuit consisting of AND and OR gates whose truth table provides outputs corresponding to the binary bits which directly relate to the decimal inputs.

As a matter of fact, every read-only memory is defined in terms of its truth table. The table shows what outputs are obtained for every given input configuration. Thus, to know the truth table is to know the function of any given read-only memory.

The diode memory of Fig. 10-4 can be manufactured in several different ways. The illustration shows a memory which might be made of discrete components, using diodes only in those locations where needed. As an alternative, the read-only memory might be manufactured as an integrated circuit containing diodes at all possible junctions. The actual programing of the read-only memory, then, would be performed by the unique metal interconnection mask for that particular integrated circuit. Then, the 1s and 0s for each particular bit location are determined by connecting or not connecting the diode in that location.

The diode memory shown in Fig. 10-4 is fine for use where only one input line at a time will be enabled. However, most ROMs are designed to produce specified outputs for every combination of inputs, where there is no restriction on the states the input lines may assume.

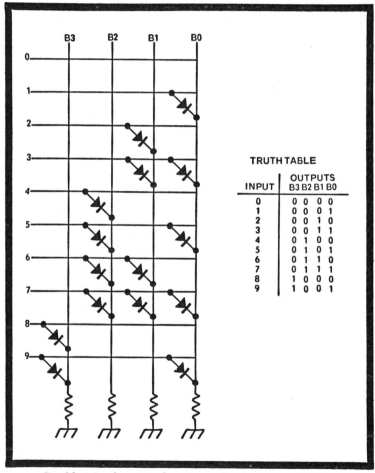

TRUTH TABLE

| INPUT | OUTPUTS B3 B2 B1 B0 | | | |
|-------|----|----|----|----|
| 0 | 0 | 0 | 0 | 0 |
| 1 | 0 | 0 | 0 | 1 |
| 2 | 0 | 0 | 1 | 0 |
| 3 | 0 | 0 | 1 | 1 |
| 4 | 0 | 1 | 0 | 0 |
| 5 | 0 | 1 | 0 | 1 |
| 6 | 0 | 1 | 1 | 0 |
| 7 | 0 | 1 | 1 | 1 |
| 8 | 1 | 0 | 0 | 0 |
| 9 | 1 | 0 | 0 | 1 |

Fig. 10-4. Diode-matrix-forming read-only memory (ROM).

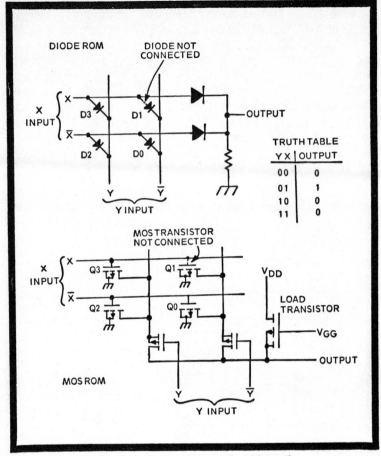

Fig. 10-5. Simple ROM, fully addressed.

A very simple example of such an ROM is shown in Fig. 10-5. This ROM has two address variables, X and Y. For each combination of X and Y, there is a defined output; either a 1 or a 0. Where a 1 output is desired, the diode or transistor at the appropriate junction is not connected. Where the connection is made, a 0 output is obtained for the state which corresponds to that intersection. The ROM shown is trivial, because it is clear that a two-input gate could perform the same function. But, conceptually, the same idea is applied to ROMs with many input variables and more than one output.

There are two versions of the same functional ROM shown, to illustrate that MOS transistors as well as bipolar elements can perform the desired memory function. Note that both versions are of the type normally manufactured as integrated circuits. That is, the 1s and 0s are programed into the ROM by connecting or not connecting the individual logic diodes or transistors.

As a final example of the type of memory obtained with the ROM arrangement, Fig. 10-6 shows a complex read-only memory array in block diagram form. This array has 16 different address locations for each of 4 output lines. Hence, it is sometimes referred to as a 64-bit ROM. This particular memory is connected to perform a binary-to-gray code conversion. The actual bit connections are represented in the figure by a dot at the appropriate intersection for a logic 1 and no dot for a logic 0. The versatility of ROMs is that, given a number of input variables, the output lines can be programed to give any conceivable pattern of 1s and 0s.

### Programmable Read-Only Memories

Read-only memories, as purchased from an integrated-circuit manufacturer, are normally already preprogramed to perform some standard table lookup function. If a large

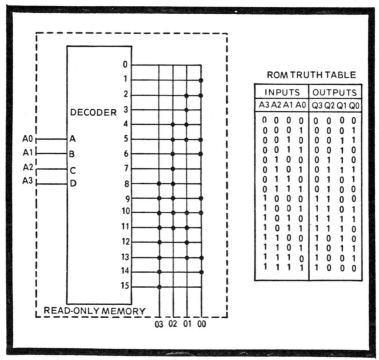

| INPUTS | | | | OUTPUTS | | | |
|---|---|---|---|---|---|---|---|
| A3 | A2 | A1 | A0 | Q3 | Q2 | Q1 | Q0 |
| 0 | 0 | 0 | 0 | 0 | 0 | 0 | 0 |
| 0 | 0 | 0 | 1 | 0 | 0 | 0 | 1 |
| 0 | 0 | 1 | 0 | 0 | 0 | 1 | 1 |
| 0 | 0 | 1 | 1 | 0 | 0 | 1 | 0 |
| 0 | 1 | 0 | 0 | 0 | 1 | 1 | 0 |
| 0 | 1 | 0 | 1 | 0 | 1 | 1 | 1 |
| 0 | 1 | 1 | 0 | 0 | 1 | 0 | 1 |
| 0 | 1 | 1 | 1 | 0 | 1 | 0 | 0 |
| 1 | 0 | 0 | 0 | 1 | 1 | 0 | 0 |
| 1 | 0 | 0 | 1 | 1 | 1 | 0 | 1 |
| 1 | 0 | 1 | 0 | 1 | 1 | 1 | 1 |
| 1 | 0 | 1 | 1 | 1 | 1 | 1 | 0 |
| 1 | 1 | 0 | 0 | 1 | 0 | 1 | 0 |
| 1 | 1 | 0 | 1 | 1 | 0 | 1 | 1 |
| 1 | 1 | 1 | 0 | 1 | 0 | 0 | 1 |
| 1 | 1 | 1 | 1 | 1 | 0 | 0 | 0 |

Fig. 10-6. Binary-to-gray code conversion using ROM.

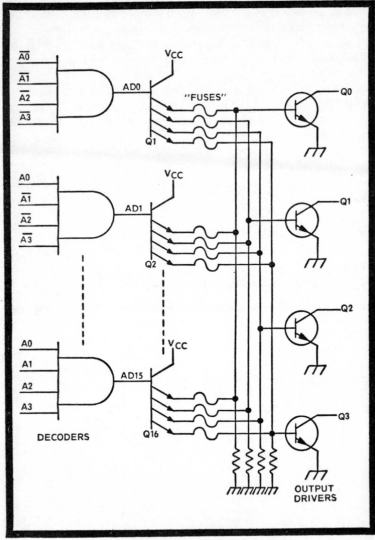

**Fig. 10-7. Bipolar programmable ROM.**

quantity of a particular memory configuration is desired, the manufacturer will fabricate the unit to the customer's requirements. However, a given digital system may require some unique table lookup which is not an off-the-shelf product and only one or two of the memories are needed. In this case, it is not economically feasible for the manufacturer to tool up for the necessary production cycle. As an alternative, programmable read-only memories can be purchased which have all of their programing interconnections made with a fusible material.

To program such a read-only memory, a one-time write process is performed which essentially open-circuits all of the diodes where no connection is desired. In this way, the user can make any memory function needed. Programmable read-only memories are frequently found in one-of-a-kind digital systems.

An example of a programmable read-only memory is shown in Fig. 10-7. This memory is a bipolar memory as opposed to the MOS memory, because it uses bipolar transistors. A programmable read-only memory can just as easily be made in the MOS version, but its operating speed is typically somewhat slower than the bipolar version.

Referring to the figure, the 16 addresses (AD0 through AD15) are decoded in a standard logic fashion to apply an enable voltage to one of 16 multiemitter transistors with "fuses" connecting the emitters to the output driver stages. There are four outputs, labeled 00 through 04, each output requiring a separate driver transistor. To program this memory, the desired address is selected, then a fairly large negative collector voltage is applied to the output being programed. As a result, the transistor base—collector junction becomes forward biased and draws enough current to cause the selected "fuse" to open, thereby programing a logic 1 in that location.If the fuse is left unblown, a logic zero is retained at the location.

**Read-Mostly Memories**

There are certain instances when a particular set of lookup tables may be needed for weeks or months. This is definitely an application for a read-only memory. Occasionally, however, the user would like to modify the data in these tables to meet his changing needs. Here, a **read—write** memory would be preferred; but a standard read—write memory cannot be used: it uses flip-flops for data storage; and flip-flops lose their stored information when power is removed and reapplied.

One solution to the problem is a read-mostly memory (RMM). In the logic circuit this type of memory is used exactly like a read-only memory. The difference is that the read-mostly memory can be repeatedly reprogramed to give new sets of outputs. Instead of the fusible material in programmable read-only memories, the read-mostly memory uses transistors with metal nitride—oxide (MNOS) as the semiconductor element for storage of data bit patterns. Storage is accomplished by

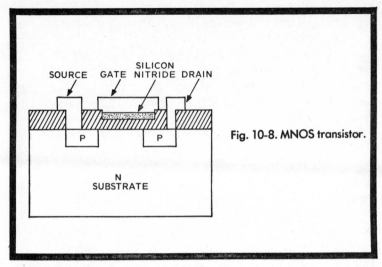

Fig. 10-8. MNOS transistor.

electrically altering the threshold voltages of the individual transistors that make up the memory.

Physically, the MNOS transistor like the one shown in Fig. 10-8 is quite similar to an ordinary MOS transistor. The mechanism for varying its threshold voltage lies in the ability to trap electrical charges in the silicon nitride gate insulator. If a positive charge is trapped, the threshold voltage of the transistor is lowered. Similarly, if a negative charge is trapped, the threshold is raised. A low-threshold transistor (one which will turn on) represents a logic 1, while a high threshold transistor (one which remains off) represents a logic 0.

When a number of MNOS transistors are combined into a memory array, the result of the trapped charges is to store logic 1s and 0s in a unique pattern to form a custom-programed read-only memory. The terminology "read-mostly memory" is derived from the fact that, at least occasionally, the memory must be written into. But, for the most part, only the read function is performed.

The read-mostly memory is nonvolatile in that power can be removed from the transistors then reapplied as often as desired while the trapped charges representing the binary data will not be lost. After a very long period, perhaps as much as a year, the MNOS transistors tend to lose their charges; and therein lies their principal disadvantage. An MNOS memory array must be occasionally refreshed (reprogramed) so that it will not lose the trapped charges.

## Random-Access Memories

A *random access memory* (RAM) is a read—write memory consisting of a number of memory cells which can be randomly accessed, based upon an input address. For large-scale integration, the memory cells are usually cross-coupled multiemitter transistor flip-flops with appropriate input and output select logic added to perform the random-access function. This random-access memory is a volatile type, because all data is lost if power is removed.

A schematic diagram of a basic RAM cell is shown in Fig. 10-9. To understand the operation, assume initially that the flip-flop is in some given state and that the address select line is low. The two transistors in this case look exactly like a standard common-emitter flip-flop with the emitters tied to ground. Clearly, this flip-flop will store a logic 1 or a logic 0 indefinitely, so long as power is not removed.

Next, assume that at some later time the address select line is set high. Then, the two output emitters Q and $\bar{Q}$ will independently have current flow or not, depending on which transistor is conducting and which is cut off. Thus, to read a bit from the flip-flop, it is only necessary to sense which emitter has current flow to determine the state of the flip-flop.

Finally, if it is desired to write new data into the flip-flop, the address select line is again set high so that the two emitters are independent. Now, if a logic 0 is placed on one emitter and a logic

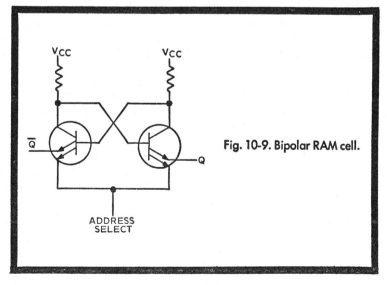

Fig. 10-9. Bipolar RAM cell.

1 on the other, the transistor with the lower emitter voltage will conduct and the opposite transistor will cut off. Hence, a logic 0 will be stored by the transistor which has a logic 0 applied to its emitter, while a logic 1 will be stored by the transistor which has a logic 1 applied to its emitter.

To understand how the individual memory cells make up a complete memory array, a block diagram for a typical random-access memory is shown in Fig. 10-10. This memory contains 64 bits of storage, arranged in memory cells numbered by row and column. There are 16 rows, each containing 4 memory cells. Each row can be written into or read out from individually. To write data into the memory, a **write enable** signal is required, in addition to four data bits at the data input lines. The address specified at the address input lines defines the memory cells that will receive the input data. The read function is performed by setting the write enable line low and applying an address indicating the row of memory cells to be read out.

Typical usage of a random-access memory requires two separate address generators, one for writing and one for reading. Data is written into known locations in the memory as designated by the written address generator. Hence, when data is retrieved from the memory, it is accessed by knowing in advance the memory location where the data resides and addressing this location. As will be shown, other memory types do not require that the specific memory address be known.

**Content Addressable Memories**

The random-access memories just described depend upon prior knowledge of memory location to read out data. A *content addressable memory* (CAM), however, reads out data by applying a known data set at the inputs then searching memory contents to see if there are any locations containing the same information. If there are, a *match* condition occurs and the addresses of all matching locations are made available. Normally, the content addressable memory will be organized such that some key portion of the data will be searched, while the rest of the data is ignored.

For example, a particular memory might contain frequency, pulse width, and amplitude information concerning a given set of pulses. A standard search function might be to scan memory to see how many pulses of a given frequency there are in memory. Hence, only the frequency portion of the data would be searched for a match condition. Once a match condition is obtained, the specific memory locations are known, and the entire data files can then be read out by direct memory addressing.

Writing into the memory is also done by direct address. The determination of a write address is based either on known empty memory locations or by searching for locations in which data is no longer needed, then writing in those addresses.

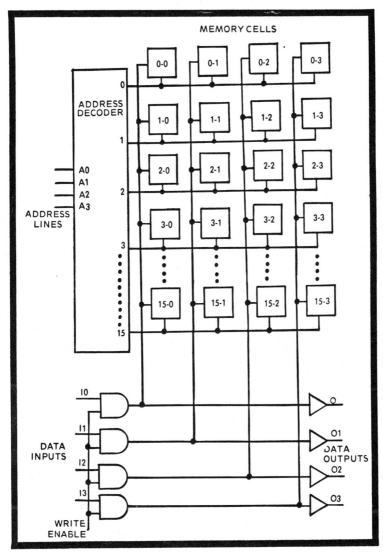

Fig. 10-10. Typical random-access memory organization.

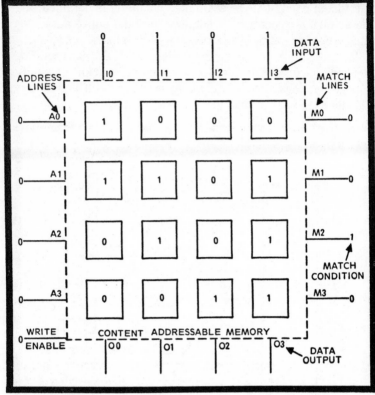

Fig. 10-11. Search operation in content addressable memory.

A simplified diagram of a 16-bit content addressable memory is shown in Fig. 10-11. There are four rows in the memory, each containing four memory cells. Each row has an individual address bit, allowing the four memory cells to be searched, read, or written into.

Assume that the memory contains the data bits shown, and that a search operation is to be performed to determine if the number five (0101) is contained anywhere in the memory. If the address bits are all set to 0, and the search data (0101) is applied at the data input lines, a match condition will occur at all rows which contain the desired information. In this case, only row three contains five. If it is desired, row three can be read out by simply setting address bit $\overline{A}2$ to a logical 1. To write in the memory, the write enable line is set to a logic 1 along with the selected address line. The data present at the data input lines will then be written into the selected row of the memory.

**First-In—First-Out Memories**

A modification of the random-access memory concept is the idea of a **first-in—first-out** (FIFO) memory. Also referred to as a **silo** memory, the analogy is quite appropriate. In a farmer's silo, the feed is put into the top (usually at quite a high speed) and is taken out more slowly at the bottom. The first feed into the silo goes directly to the bottom of the silo, thus also becoming the first feed out. The first-in—first-out memory receives data at one data rate and outputs data at some other unrelated data rate. The first data in, like the silo, becomes the first data out. A typical use for a first-in—first-out memory might be the interface between a high-speed digital computer and a much slower typewriter. Blocks of data can be fed into the first-in—first-out memory at high speed, until the memory is full. Then, the typewriter can read the same data out at a slower rate. When the memory is empty, the computer can again refill the memory cells at a high data rate.

A block diagram of a first-in—first-out memory is shown in Fig. 10-12. Recalling the random-access memory operation, separate read address and write address generators were commonly used to access specific data locations. With the first-in—first-out memory, the read and write addresses each increment sequentially so that two counters advances one count each time data is entered into the memory while the read counter advances a single count each time data is read out. Notice that, external to the memory, there is no need to know memory locations. Indeed, there is no provision to access data by specific location, even if this were desired.

## ADDERS

Adder circuits such as the parallel full adder described in Chapter 9 are naturals for medium-scale integration. Consisting entirely of simple gate circuits, a typical 4-bit full adder can be included in a single package. The main limiting factor with MSI adders is the number of input and output leads required. In the case of a 4-bit full adder, there are 8 inputs for the bits to be added, 4 outputs representing the sum bits, a carry input, and a carry output. For this simple circuit, there are 14 input and output signal leads required, not including power and ground.

## SUBTRACTORS

Subtractors, like full adders, are easy to implement as medium-scale integrated circuits. Actually, a subtractor is

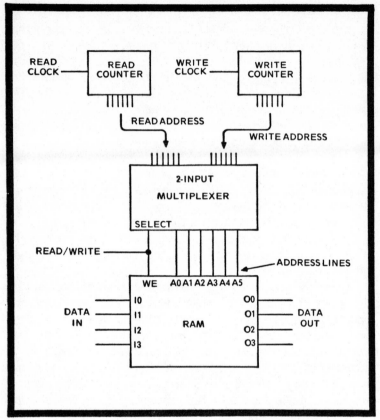

**Fig. 10-12. First-in—first-out memory block diagram.**

usually built from a full adder by adding inverters on the inputs to the number being subtracted. As was mentioned in Chapter 1, *two's complement* notation is extremely useful for addition and subtraction of binary numbers. Use of the full adder as a subtractor is a good example of this concept.

Recalling, *two's complement* subtraction was performed by inverting all of the bits in the number being subtracted, adding one, then adding the two numbers together. This is exactly what is done in Fig. 10-13. Although the inverters are shown external to the full adder, a subtractor package would include the inverters inside the unit. Several examples are shown to demonstrate the *two's complement* technique of subtraction. These examples show positive numbers being subtracted from one another; however, the circuit works equally well for negative numbers, provided that they are in *two's complement* notation.

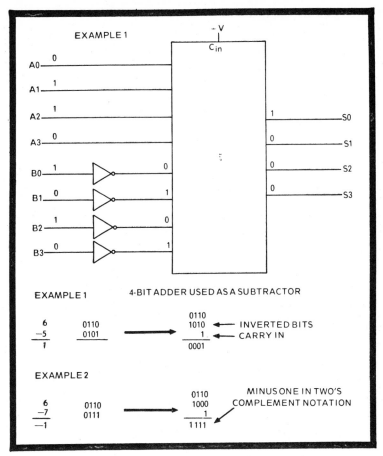

Fig. 10-13. Subtractor using two's complement arithmetic.

## COMPARATORS

A digital comparator circuit is merely an extension of the use of adder and subtractor circuits. Assuming there are two numbers, A and B, to be compared, the technique consists of subtracting B from A, then noting whether the result is positive, negative, or zero. If A − B is positive, then A is greater than B. Similarly, if A − B is negative, then B is greater than A. The condition where A − B = 0 is decoded to show that A = B.

The comparator circuit has the same type of limitations as adders and subtractors. The number of bits which can be compared in a single package is strictly a function of the number of leads which can be made available in a given package configuration.

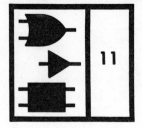

# 11 Digital Interface Circuits

The dictionary definition of an interface is: *A common boundary between two parts of matter or space*. In the case of digital circuits, the "two parts of matter or space" can be interpreted to mean two physical or functional electronic units. Hence, an interface circuit could be one which transmits and receives digital data, say between two separate electronic chassis; or it might be one which converts analog information to digital data, or vice versa. This chapter explores some of the more common types of interface circuits likely to be encountered.

## LINE DRIVERS AND RECEIVERS

Line drivers and receivers are the amplifier circuits which transfer digital data between units. Whenever such a data transfer occurs, problems of noise entering the system are also likely to occur. The noise may be ground noise between two chassis, or it may be crosstalk from signals inductively coupled between wires in cable harnesses. Whatever the cause, the basic solution to the noise problem lies in improving the signal-to-noise ratio at the receiving end. Three different approaches to improving signal-to-noise ratio will be considered.

### Level Shifters

Probably the easiest way of improving signal-to-noise ratio is simply to increase the amplitude of the signal at the source. Given that the amount of noise on a particular transmission line remains constant, then the improvement in signal-to-noise ratio is in direct proportion to the increase in signal amplitude. Level shifters are the digital circuits which perform this function. A simple level shifter might consist of a common-emitter transistor amplifier, which receives signal at DTL or TTL logic levels and provides output voltage swings of +12V and −12V, to represent the logic 1 and logic 0 states, respectively. Disadvantages to this technique are the requirement for higher-voltage power supplies and the increased power consumption inherent in such a circuit.

## Coaxial Cable Drivers

Another method for improving signal-to-noise ratio is by decreasing the amount of noise on the lines. This time, assuming that the **signal** amplitude remains constant, then the signal-to-noise ratio improvement is directly related to the reduction in **noise** at the receiving end. The most common method for noise reduction is through the use of properly terminated shielded cable, driven by low-impedance driver circuits. Here, coax drivers are the digital circuits which perform the digital data transmission. A straightforward example of a coax driver circuit is an emitter-follower transistor amplifier which drives a 50-ohm coax cable terminated at the receiving end with a 50-ohm resistor. The main disadvantage to this method is that shielded cable is quite expensive and system costs rise rapidly if large amounts of cable are required.

## Differential Line Drivers and Receivers

Finally, a third method for improving signal-to-noise ratio is through differential amplifier techniques, utilizing the improvements gained through common-mode versus differential-mode gains. Differential line drivers and receivers are the digital circuits that perform this function. In general, if two wires are kept physically close to one another, as in a twisted pair, extraneous signals induced into one wire will also be induced into the other. Such signals, which occur simultaneously in two wires, are called common-mode signals. Similarly, any ground noise in the two wires will also tend to be common-mode noise. If, on the other hand, the signals on the two wires swing in opposite directions, the signals are said to be differential-mode signals.

Differential-mode signals can be purposely generated, as shown in Fig. 11-1. The line driver which generates the differential mode signal is simply an AND gate and a NAND gate, with parallel inputs driving two separate wires. The input signal will appear with positive polarity on one wire and with inverted polarity on the other wire. The key to differential-mode operation is in the line receiver circuit.

The line receiver consists of a basic differential amplifier circuit. To understand circuit operation, consider first the case where a positive signal is present at the base of Q1, while the base

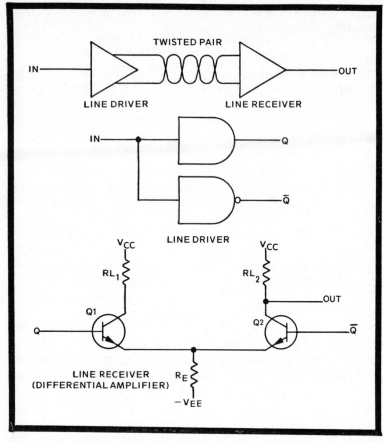

Fig. 11-1. Differential line driver—receiver circuit.

of Q2 is at ground potential. This is a typical differential-mode signal. Since the base of Q1 is positive, Q1 will conduct heavily, causing current to increase through resistor $R_E$. With the base of Q2 held at ground, this represents a reverse bias to transistor Q2, and current flow through $R_E$ decreases by an amount equal to the increase caused by Q1. The result is that there is no net change in the current flow through $R_E$; thus, the emitters of both transistors can be considered as being at virtual ground. In this mode, the gain of the amplifier is quite high.

For common-mode operation, assume that the two bases are tied together, as is functionally the case if both base voltages change by the same amount and in the same direction. A positive voltage at the base of Q1 will cause an increase in current through $R_E$, but a similar increase at the base of Q2 also causes the same kind of current increase. To both transistors, the resultant positive voltage at their emitters is a negative feedback, causing a decrease in conduction through both transistors. Of course, this decrease in conduction offsets the increase caused by the base going more positive, thus very little signal gain is achieved.

What has been shown is that differential-mode signals (the true signals of interest) receive very high gain from the differential amplifier, while common-mode signals (noise) receive very low gain. The net result is a significant improvement in signal-to-noise ratio.

## DIGITAL-TO-ANALOG CONVERTERS

Often, information which is in digital format is required in analog form. When such a conversion is needed, a digital-to-analog converter is used. An example of a requirement for such a conversion is the situation where a number of parallel digital data bits form audio or video information. In order to display the information on an oscilloscope or listen to the audio, the digital data must first be converted to an analog waveform. Typically, digital-to-analog conversion is performed with a resistive ladder network followed by an operational amplifier. Since each of these functions is somewhat unique, they are described separately below.

### Resistive Ladder Networks

An extremely simple method of converting binary digits to analog voltages recognizes the basic characteristics of binary data. There are only two voltage levels present for any binary number; thus, all bits of the number have the same voltage value when in their logic 1 states, and the same voltage value when in their logic 0 states. If a resistive network can be constructed which gives a large voltage output for the most significant bits of the number and smaller voltages for the least significant bits, then a digital-to-analog conversion will have been performed. Resistive networks which are scaled to perform this function are called resistive ladders or ladder adders. The latter name is derived from the fact that the network sums or adds the voltage contributions from the various bits to form the resultant analog voltage.

One very common type of resistive ladder is the binary ladder shown in Fig. 11-2. This ladder uses resistors which are

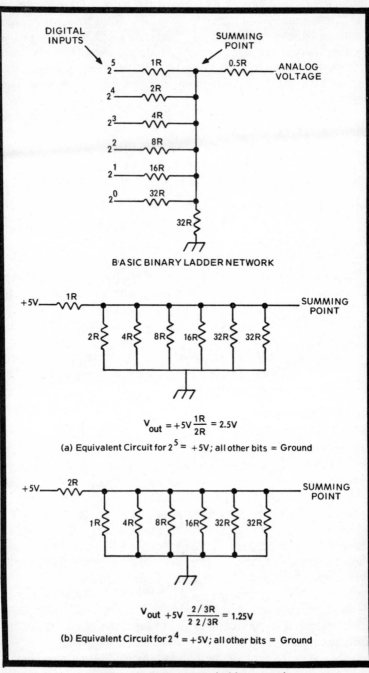

$$V_{out} = +5V \frac{1R}{2R} = 2.5V$$

(a) Equivalent Circuit for $2^5 = +5V$; all other bits = Ground

$$V_{out} +5V \frac{2/3R}{2\ 2/3R} = 1.25V$$

(b) Equivalent Circuit for $2^4 = +5V$; all other bits = Ground

**Fig. 11-2. Binary resistive ladder network.**

selected in binary increments to form a voltage-divider network. Hence, the most significant bit uses a value of 1R, the next bit 2R, etc. The network is terminated with a resistor equal to the least-significant-bit resistor.

To understand the functioning of this ladder, assume that the logic levels happen to be +5V for a logic 1 and ground for a logic 0. Also, assume that initially only the $2^5$ bit is at a logic 1 and that all other bits are logic 0. A voltage divider is formed, consisting of one series resistor and the remaining resistors in parallel to ground. According to the rules for summing parallel resistances, it is seen that 32R in parallel with 32R becomes 16R, and 16R in parallel with 16R sums to 8R, and so on. The resultant total resistance to ground from the summing point is 1R. Obviously, if +5V is divided between two resistors of value 1R, the point in the middle is at +2.5V

Now, assume that only the $2^4$ bit is at a logic 1 and the rest go to logic 0. Here, the parallel resistors collapse to 2/3R, producing an output of 1.25V. This is one-half the voltage obtained when the $2^5$ bit was a logic 1. From the above, it can be seen that each bit produces one-half the voltage of the next significant bit. The summing point becomes an analog voltage, proportional to the sum of all the individual voltages applied by the resistors.

One disadvantage to the binary ladder is that many different values of resistors are required, and also that each digital stage is terminated in a different resistance value. A scheme which uses uniform resistor values is the R−2R ladder shown in Fig. 11-3.

Assume that the $2^5$ bit is a logic 1 and that all other bits are logic 0. Starting at the bottom of the ladder, it is seen that 2R to ground in parallel with 2R to the $2^0$ bit (also ground) sums to 1R. But 1R in series with 1R is 2R, and this is in parallel with 2R to ground for the $2^1$ bit. Continuing the summing, the network collapses to a voltage divider consisting of two resistors of value 2R. Thus, when the $2^5$ bit is a logic 1, the output of the ladder is +2.5V, just as with the binary ladder. Considering the case where the $2^4$ bit is a logic 1 and all other bits are logic 0, it is seen that the network collapses to a simple divider which this time produces an output of +1.25V. These results are identical with those obtained from the binary ladder network.

## Operational Amplifiers

The resistive ladder networks previously described work well for driving very high-impedance circuits. However, if an attempt is

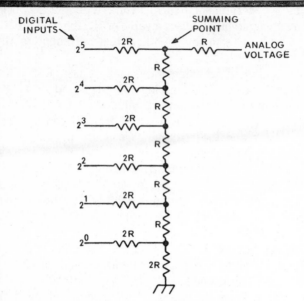

BASIC R—2R LADDER NETWORK

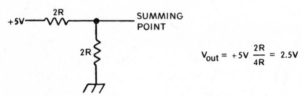

$$V_{out} = +5V \frac{2R}{4R} = 2.5V$$

**(a) Equivalent circuit for $2^5 = +5V$; All other bits = ground.**

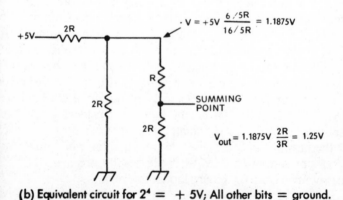

$$V = +5V \frac{6/5R}{16/5R} = 1.1875V$$

$$V_{out} = 1.1875V \frac{2R}{3R} = 1.25V$$

**(b) Equivalent circuit for $2^4 = +5V$; All other bits = ground.**

Fig. 11-3. R—2R resistive ladder network.

made to drive a low-impedance circuit, there is insufficient drive from the flip-flops which supply voltages to the individual resistors and a degradation of voltage levels occurs.

Uniform known voltage levels are a prerequisite for proper operation of the resistive ladder network. Operational amplifiers provide the necessary impedance matching and buffering as well as amplification of the analog voltage swings to achieve the well behaved voltage levels.

Basically, an operational amplifier consists of a differential amplifier quite similar to the one described previously (and shown in Fig. 11-1) followed by several stages of level shifting and additional voltage gain. The characteristics which distinguish an operational amplifier are (1) very high input impedance, (2) very low output impedance, and (3) a very large voltage gain. The high input impedance assures that the amplifier won't load down the circuit driving it, while the low output impedance provides good drive capability to other circuits. By having a large voltage gain, negative feedback can be employed (usually in the form of a feedback resistance, $R_F$) to control the overall gain of the amplifier in a particular application.

Some of the most common operational-amplifier configurations are shown in Fig. 11-4. Note that each configuration has a plus input and a minus input, corresponding to the two inputs of a differential amplifier.

**Inverting Amplifier**. Probably the most used configuration is the standard inverting amplifier shown at the top of the figure. Feedback resistor $R_F$, in conjunction with the input resistance, $R_i$, forms the negative feedback path which determines the gain of the circuit. For this configuration, the voltage gain is defined as the ratio $R_F/R_i$. Thus, for example, when $R_F = R_i$, then the gain of the circuit is unity. Similarly, if $R_F = 10K$ and $R_i = 5K$, the gain of the circuit is two. In other words, for an input signal amplitude of 2V, an output amplitude of 4V will be obtained.

**Noninverting Amplifier.** The noninverting amplifier configuration is quite similar to the inverting amplifier, except that it uses the other input to the differential amplifier. The gain of this circuit is $1 + R_F/R_i$. A very common setup here is the special case where a unity-gain voltage follower is desired. From the equation, it is seen that in order to obtain unity gain, $R_F$ should equal zero and $R_i$ should equal infinity. This is equivalent to saying that $R_i$ does not exist and $R_F$ is a short circuit. Under these conditions, the circuit shown in Fig. 11-5 results. The unity-gain voltage follower is often used where there is no change in signal

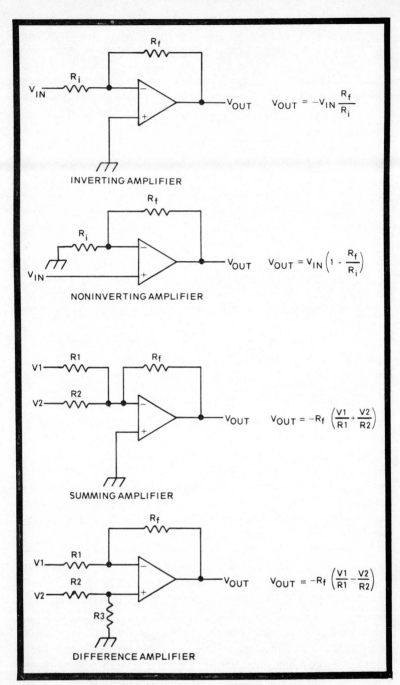

Fig. 11-4. Typical operational amplifier configurations.

168

level desired, but buffering is required. As previously noted, the operational amplifier's high input impedance does not load the input circuit, while the low output impedance provides excellent drive capability.

**Summing and Difference Amplifiers.** Referring again to Fig. 11-4, the summing amplifier configuration is seen to be a special case of the inverting amplifier. Of particular interest here is the fact that the summing amplifier input circuit looks very much like the two **ladder adder** networks previously described. Indeed, when the ladder adder network and operational amplifier are combined, the resulting configuration is that of a digital-to-analog converter. A diagram of a complete digital-to-analog converter is shown in Fig. 11-6. Note that the operational amplifier has a ladder adder circuit at its input. Feedback is provided by a single resistor, which in this case gives unity gain. The main function of the operational amplifier in this circuit is to provide the drive necessary for the following circuits and to buffer the resistive ladder (and flip-flops driving it) from the effects of circuit loading. The resistor on the plus input of the amplifier is included so that the differential amplifier sees approximately the same impedance on both inputs.

## ANALOG-TO-DIGITAL CONVERTERS

Analog information exists in many different forms. This information may be represented by a continuously variable

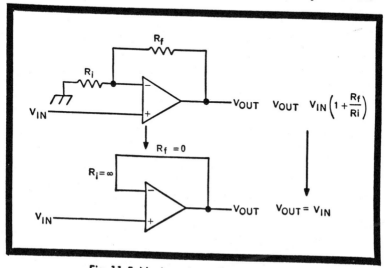

Fig. 11-5. Noninverting unity-gain amplifier.

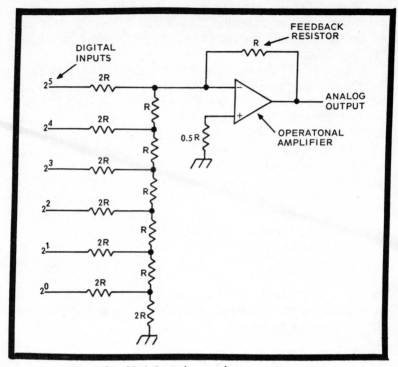

**Fig. 11-6. Digital-to-analog converter.**

voltage, or it may be represented by the position of a shaft or lever. Also, the analog action can range from quite slow, as with a shaft rotation of several revolutions per minute, to very fast, as is the case of a waveform varying at a rate of several megahertz. Because of the wide variety of analog devices, there is also a corresponding variety of analog-to-digital conversion techniques. The following paragraphs will not attempt to explore all of the known techniques, but rather will describe the basics of several commonly used analog-to-digital converters.

**Elapsed-Time Technique**

One simple way of converting an analog voltage to a digital number is by generating an analog ramp voltage and counting the number of clock periods it takes the ramp voltage to rise to a value which equals an unknown analog voltage. Providing that the ramp is linear and that its slope remains constant, the elapsed time to reach a given voltage is directly proportional to the amplitude of that voltage.

A block diagram of a converter using this technique is shown in Fig. 11-7. The top portion of the figure represents the basic converter, which will be described first.

The analog ramp generates a sawtooth waveform, starting at the most negative voltage to be encountered and graduating to the most positive voltage expected. The input voltage is compared in an analog comparator against the ramp voltage. The analog comparator is simply an operational amplifier with a zener diode for feedback. If the ramp voltage is greater than the analog input, the comparator output equals the zener voltage. If the ramp amplitude is less than the analog input, the comparator output is essentially at ground potential. The flip-flop acts as a control for the counter. Upon receipt of a start pulse, the ramp starts at its most negative value, the flip-flop is set, and clock pulses are applied to the counter. As the ramp voltage increases, so does the counter contained in the counter. When the ramp reaches the same value as the input voltage, the comparator output goes positive, resetting the flip-flop and stopping the clock

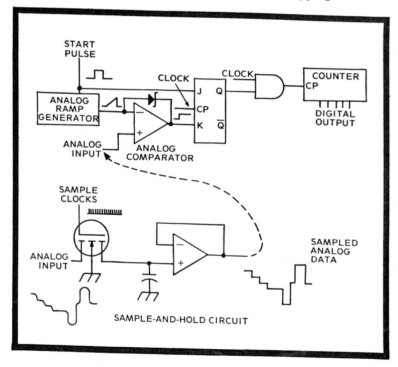

Fig. 11-7. Elapsed-time technique.

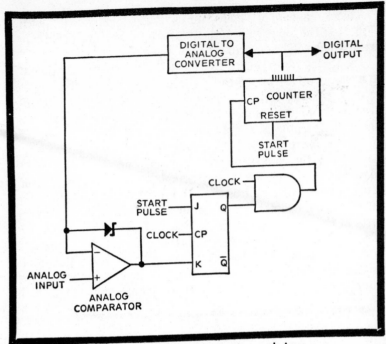

Fig. 11-8. Binary ramp comparison technique.

pulses to the counter. At this time, the state of the counter represents the analog input voltage.

If the analog voltage changes very slowly, the above circuit is sufficient. However, with a more rapidly changing analog input, the voltage may change significantly between the time the ramp is started and the time when the conversion is completed. A method for holding the analog input steady is the sample-and-hold circuit shown at the bottom of Fig. 11-7. Here, a MOS transmission gate is used to charge a capacitor connected to an operational amplifier. Each time a clock pulse occurs, the capacitor charges. Between clock pulses, the MOS gate is open and the capacitor stores a constant voltage until the next clock pulse occurs. In this way, the analog-to-digital converter is presented with a constant voltage for one entire conversion cycle; it is presented with some other constant voltage for the next conversion cycle.

### Binary Ramp Comparison Technique

One problem with the elapsed-time technique is that the generated ramp must be very smooth and linear if accurate results are to be obtained. An alternative approach is to generate a ramp by clocking a digital counter and performing a digital-to-analog conversion on the counter outputs. Now, if the ramp is compared to the analog input, the point at which the ramp equals the input voltage can again be used to stop the counter. The counter state represents the value of the analog input voltage. The primary difference is that the generation of the ramp is not dependent on time. If desired, the ramp could be increased by a small amount, stopped, then increased some more. In either case, the conversion would be identical to that obtained with a smoothly increasing ramp. Of course, it is almost always advantageous to complete a conversion cycle as quickly as possible, mainly because the analog input is liable to change over a period of time.

A block diagram of a binary ramp converter is shown in Fig. 11-8. The conversion cycle starts by resetting the counter to all zeroes and setting the clock control flip-flop, thus applying clock pulses to the counter. As the count increases, the digital-to-analog converter changes the output states of the counter to an analog ramp voltage. When the ramp voltage equals the analog input, the flip-flop is reset and the conversion cycle is complete. At this time, the state of the counter represents the magnitude of the analog input. As with the previous circuit, a sample-and-hold circuit can be used if needed to hold the analog input steady during the conversion cycle.

### Successive Approximation Technique

The binary ramp technique has one major drawback. If the maximum possible positive voltage were to be converted in a 6-bit converter, the conversion process would require $2^6 = 64$ steps to complete—one for each possible count of the binary counter. Thus, the binary ramp is extremely slow. A method which greatly reduces the conversion time is the successive approximation technique. As implied by its name, this method successively approximates the input voltage by factors of two, to very rapidly converge to the correct answer. As will be seen, instead of 64 steps, the successive approximation technique requires only six steps to perform the same 6-bit conversion.

A block diagram of a 6-bit successive approximation converter is shown in Fig. 11-9. Initially, the assumption is made that the number representing the input analog voltage is 100000. Note that this value is exactly halfway between 0V and the maximum possible voltage. The number 100000 is loaded into the trial

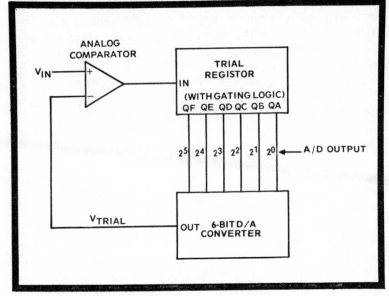

Fig. 11-9. Block diagram of successive approximation converter.

register and converted to an analog voltage. At this time, the analog comparator has two inputs, the input voltage ($V_{in}$) and the trial voltage ($V_t$).

If the input voltage is greater than the analog $V_t$ representation of 100000, the comparator output is a logic 1 and the most significant bit of the trial register remains a 1.

If the input voltage is less than the trial voltage, the comparator output is a logic 0 and the most significant bit is set to 0. For the second step, the next bit in line is set to a logic 1, while retaining the previously determined most significant bit in the trial register. This has the effect of picking a new trial voltage halfway between the remaining limits to be determined, either 110000 or 010000.

Again a comparison is made to see in which direction the input voltage lies with respect to the trial voltage. Again, if the voltage is more positive, the comparator output is a logic 1, indicating that a 1 should be in that bit position. If the input voltage is more negative than the trial voltage, the comparator output is a logic 0 and a 0 would be placed in the bit position that is now most significant. This process is repeated six times, each time halving the remaining voltage to arrive at the correct voltage more quickly, and each time writing a 1 or 0 in the appropriate bit

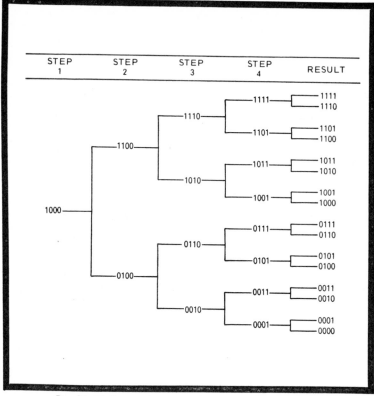

Fig. 11-10. Successive approximation trial steps.

position, based on the results of the comparison between the new trial voltage and the input voltage. Thus, the successive approximation technique performs a conversion in $n$ steps (where $n$ is the number of bits in the number) instead of in $2^n$ steps, which may be required with the binary ramp technique.

Figure 11-10 illustrates the steps which would be followed for a 4-bit conversion. At each step, one of two possible branches is picked for a new trial number, then a comparison is made to determine the actual number. The process can be extended to 5, 6, or even 10 bits, if desired. The main limiting factor is the small differential voltage which will be applied to the comparator as the trial voltage approaches the input voltage.

### Optical Shaft Encoder

The previous techniques of analog-to-digital conversion use electronic devices entirely. However, many real-world analog

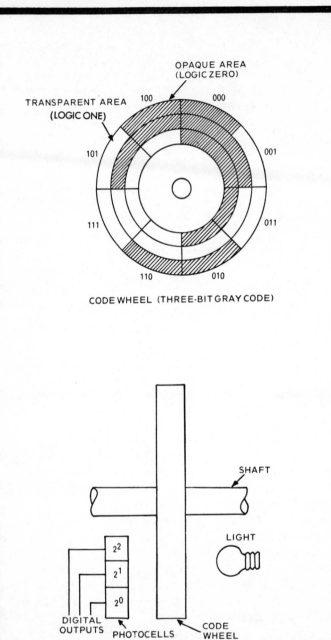

Fig. 11-11. Optical shaft encoder.

events are mechanical actions. In this case, an electromechanical device must be used to perform the conversion. A very frequently used electromechanical analog-to-digital converter is an optical shaft encoder. A simplified diagram of an encoder of this type is shown in Fig. 11-11. The encoder consists of a code wheel with opaque and transparent areas, mounted on a shaft. A light source shines through the code wheel onto three photocells, each of which emits a voltage proportional to its light inputs. If no light shines on a photocell, its output is low, representing a logic 0. If the light does shine on a photocell, that cell's output is high, representing a logic 1. By coding the wheel such that there are eight discrete otputs for an infinite variety of shaft positions, an analog-to-digital conversion has been performed. The particular code wheel in this example is constructed so that its outputs are gray-code representations of the shaft positions. Thus, if the wheel happens to be at some position between two different code outputs, the worst error which will be obtained is a single bit.

**12**

# Digital Displays

There are many occasions when digital data from various portions of a system are displayed. Sometimes the data is displayed in raw binary format, but more often it is converted to some type of decimal display so that operators can more readily interpret the results being displayed. It is not unusual for a digital display to represent the end result of an entire set of computations and logic operations. Some examples of equipments where this is the case are counters, digital voltmeters, and electronic calculators. Because digital displays are so frequently used, it is desirable to gain some familiarity with the different types of displays available. This chapter describes the most common display types encountered.

## INDIVIDUAL LAMP DISPLAYS

Almost everyone is familiar with the little red light associated with the power switch on a piece of test equipment. This light is the simplest form of a digital display. When the light is lit, the power is on (a logic 1), and when the light is dark, the power is off (a logic 0). There is no universal lamp type which is used everywhere. The type of lamp used is a function of the particular application. Some lamp types found in digital systems are incandescent, neon, fluorescent, and light-emitting diodes.

### Incandescent Lamps

An example of an incandescent lamp is the standard light bulb used in the home. These lamps operate by sending current through a very fine filament wire, causing the wire to heat up and thus emit light. Although alternating current is normally used in the home, the same principle applies if direct current is used. In general, the incandescent lamps made for digital equipment are small and can operate on low voltages. The main faults with all incandescent lamps is that their filaments tend to burn out fairly often and they waste energy because they generate heat.

## Fluorescent Lamps

An incandescent lamp gives off light due to heating of its filament. Correspondingly, it is known that an ordinary vacuum tube uses the heat from a filament in conjunction with a fairly large anode voltage to cause thermionic emission of electrons from the cathode that are attracted to the anode. Fluorescent lamps of the type used in digital equipment utilize a very similar technique. The fluorescent lamp is basically a vacuum tube diode in which the anode has been coated with a special phosphor which glows when bombarded with electrons. When the diode conducts, the lamp is on; and when the diode is off, the lamp is off. Several fluorescent lamps are shown in Fig. 12-1. The basic lamp schematic simply shows the expected diode structure. This basic lamp configuration is not frequently found. The more common fluorescent display is the numeric display. Here, the numbers are formed by placing many anode segments in a single envelope, then applying separate anode voltages to each as necessary to turn on segments which form the desired number. Typically, fluorescent displays are greenish-blue in color.

## Neon Lamps

In direct contrast to the incandescent and fluorescent lamps which utilize hot cathodes for thermionic emission, neon lamps are often referred to as cold-cathode or gas discharge displays.

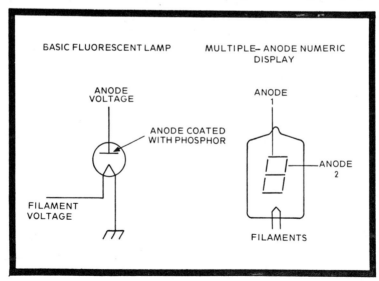

Fig. 12-1. Fluorescent displays.

Fig. 12-2. Mask for seven neon lamps.

The reason for this is in their basic functional mechanism: A neon bulb consists of a glass envelope filled with neon gas and containing an anode and a cathode. There is no filament. When a high voltage is applied between the cathode and anode terminals, the neon gas in the tube ionizes, thus causing an orange glow. Normally, neon lamps are quite tiny and are well suited for behind-the-panel displays. One typical use of this type is shown in Fig. 12-2. Seven individual neon lamps are placed behind a mask consisting of an opaque material with slots cut out where the lamps can show through. By turning on various combinations of neon bulbs, any number from 0 through 9 can be formed. One disadvantage to neon lamps is the high ionization voltage required.

### Light-Emitting Diodes

A light-emitting diode (LED) is simply a semiconductor diode which is made from a material such as gallium phosphide or gallium arsenide phosphide instead of the usual silicon or germanium normally employed for diodes. As might be guessed from the phosphide name of the material, this material emits light when properly stimulated. Actually, almost any *pn* junction emits visible light, but the efficiency of the radiation to input energy is typically quite low; also, the concentration of radiated energy may sometimes fall outside the visible light spectrum.

For reference purposes, Fig. 12-3 shows the visible light range in terms of the response curve for the human eye to various wavelengths of radiation. Notice that the color green is near the peak of the curve, while red is much harder for the eye to detect. Hence, it would seem that all light-emitting diodes would emit green light. In reality, the efficiency of green emitting diodes is so low compared to red diodes that the great majority of all light-emitting diodes are red.

The physical basis for radiation in a semiconductor diode is the recombination of holes and electrons which occurs in the depletion region and at the *pn* junction whenever a diode is forward biased. Referring to Fig. 12-4, it is seen that when a diode is biased such that the *p* side of the diode is more positive than the *n* side, the diode is considered forward biased and a large current will flow. This is basic to the operation of all semiconductor diodes. Further, the resulting current flow can be considered to consist of electrons traveling from the negative battery terminal to the positive terminal, and concurrently, it can be considered as a flow of holes from the positive terminal to the negative battery terminal. For the moment, consider only electron current flow.

The large concentration of electrons flowing through the junction is going to result in an excess of electrons also flowing through the depleted *n* material in the immediate area of the junction. Some of the electrons will therefore tend to recombine with holes in the depletion region of the *n* material in an attempt to overcome the space-charge effect in that area.

Similarly, since a hole current of equal magnitude must flow in the opposite direction, the same type of recombination will occur in the depletion region at the *p* side of the junction. From

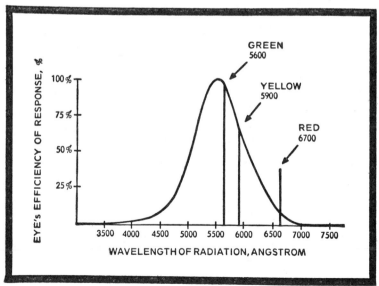

Fig. 12-3. Eye's response to radiation.

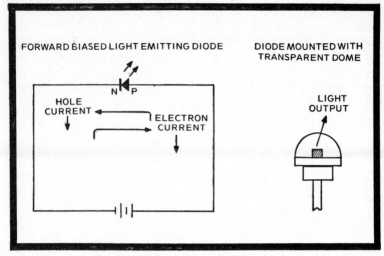

FORWARD BIASED LIGHT EMITTING DIODE

DIODE MOUNTED WITH TRANSPARENT DOME

Fig. 12-4. Light-emitting-diode mechanism.

physics, it is known that when recombination does occur, photon emission (the generation of light) also occurs. This emission or radiation is the output that is seen from a light-emitting diode. Through variation of the ratios of gallium arsenide and gallium phosphide in the $p$ and $n$ materials, the wavelength of the emission can be made to vary from approximately 5500 to 9000 angstroms. From Fig. 12-3, this is seen to range from green light to well beyond the visible red light spectrum. Other materials can be chosen to emit radiation far into the infrared portion of the spectrum if desired.

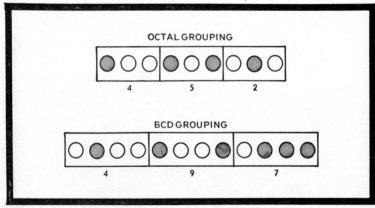

Fig. 12-5. Grouping of individual lamps.

## GROUPED BINARY DISPLAYS

A long row of lights displaying binary information can be quite a confusing sight and the pattern of *off* and *on* lights is virtually impossible to remember. Yet, some means must be provided for interpreting raw binary data. The method normally used is an octal or BCD grouping, as shown in Fig. 12-5. If the data is in straight binary format, a grouping of each three bits allows reading octal numbers directly from the binary. Note that in the example, $452_8$ is much easier to remember that $100101010_2$. If the data happens to be in BCD format, then 4-bit groupings provide a natural readout for each of the BCD decades.

## MULTIPLANE DISPLAYS

One of the first digital displays to provide data in a readable decimal format was the multiplane neon tube. This tube is a neon-filled gas envelope, with a single anode and 10 different cathodes. Each cathode is shaped to form one decimal digit, with the digits stacked one behind another. The plane which is ionized glows, thereby displaying the digit formed by the cathode. A multiplane display is shown in Fig. 12-6. The BCD-to-decimal decoder assures that only one plane will be energized at any given time.

## SEVEN-SEGMENT DISPLAYS

An improvement over multiplane displays is the seven-segment display shown in Fig. 12-7. This display shows any desired digit on a single plane. The display is made up of 7 segments which can be illuminated in various combinations to form any one of the 10 decimal digits. Usually, the seven-segment codes are decoded from BCD according to the truth table shown. There may be a slight variation in some applications, depending on whether the six and nine digits include a tail. For example, if a tail were included on the digit six, the "a" segment (top horizontal bar) would also be illuminated. Seven-segment displays are found using any of the previously mentioned lamp types.

## LIGHT-EMITTING-DIODE ARRAYS

The seven-segment display is quite good, but the decimal digits are slightly crude; and if one segment happens to burn out, an incorrect display can result. For example, if the "g" segment fails, an 8 will be displayed as a 0. An even more realistic set of

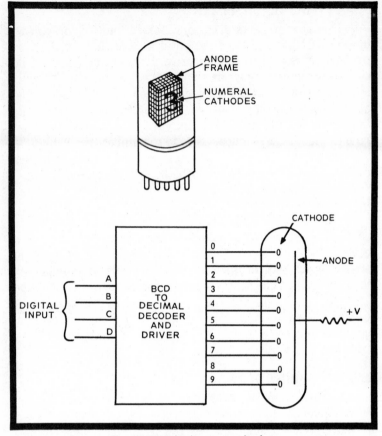

Fig. 12-6. Multiplane neon display.

numerals is possible by using an array of light-emitting diodes formed into a matrix consisting of several rows and columns. A typical $5 \times 7$ matrix is shown in Fig. 12-8. This matrix has a noticeable advantage: not only do the numerals appear better formed, but if one diode in the array should fail, the displayed number is still quite readable. The entire array is usually contained in the same integrated-circuit package as its decoder. Therefore, a light-emitting-diode array is normally addressed with a 4-bit BCD code and no external decoding is necessary.

## SCANNED ALPHANUMERIC DISPLAYS

The light-emitting-diode array concept can easily be expanded to include display of letters as well as numbers. Such a readout is called an **alphanumeric display**. Since a simple BCD

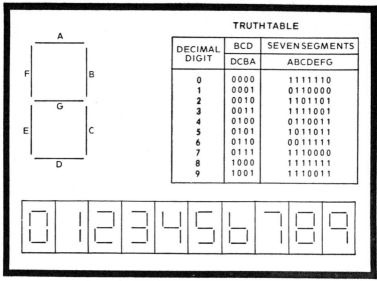

Fig. 12-7. Seven-segment display.

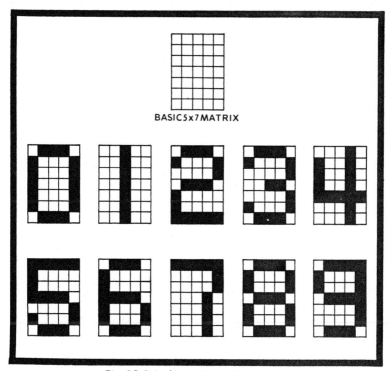

Fig. 12-8. Light-emitting-diode array.

decoder will not suffice for the large quantity of characters possible, this type of array is typically driven by a read-only memory addressed by a binary code. The binary code must contain 10 combinations for the 10 decimal digits, 26 combinations for the letters of the alphabet, and enough other combinations to display such special characters as periods, question marks, exclamation points, colons, and the like. Since at least 36 combinations are required, the smallest number of binary bits which can be used is 6. One commonly used 6-bit alphanumeric code is called ASCII (American Standard Code for Information Interchange). Table 12-1 shows the ASCII code combinations and the characters they represent. The function of the read-only memory is to convert the binary code bits to the proper combinations necessary to drive the 5 × 7 array.

Any meaningful alphanumeric display will consist of several characters displayed simultaneously. Each character displayed

**Table 12-1. Six-Bit ASCII Code.**

| A6 A5 A4 A3 A2 A1 | CHARACTER | A6 A5 A4 A3 A2 A1 | CHARACTER |
|---|---|---|---|
| 0 0 0 0 0 0 | @ | 1 0 0 0 0 0 | BLANK |
| 0 0 0 0 0 1 | A | 1 0 0 0 0 1 | ! |
| 0 0 0 0 1 0 | B | 1 0 0 0 1 0 | " |
| 0 0 0 0 1 1 | C | 1 0 0 0 1 1 | # |
| 0 0 0 1 0 0 | D | 1 0 0 1 0 0 | $ |
| 0 0 0 1 0 1 | E | 1 0 0 1 0 1 | % |
| 0 0 0 1 1 0 | F | 1 0 0 1 1 0 | & |
| 0 0 0 1 1 1 | G | 1 0 0 1 1 1 | ' |
| 0 0 1 0 0 0 | H | 1 0 1 0 0 0 | ( |
| 0 0 1 0 0 1 | I | 1 0 1 0 0 1 | ) |
| 0 0 1 0 1 0 | J | 1 0 1 0 1 0 | * |
| 0 0 1 0 1 1 | K | 1 0 1 0 1 1 | + |
| 0 0 1 1 0 0 | L | 1 0 1 1 0 0 | , |
| 0 0 1 1 0 1 | M | 1 0 1 1 0 1 | - |
| 0 0 1 1 1 0 | N | 1 0 1 1 1 0 | . |
| 0 0 1 1 1 1 | O | 1 0 1 1 1 1 | / |
| 0 1 0 0 0 0 | P | 1 1 0 0 0 0 | 0 |
| 0 1 0 0 0 1 | Q | 1 1 0 0 0 1 | 1 |
| 0 1 0 0 1 0 | R | 1 1 0 0 1 0 | 2 |
| 0 1 0 0 1 1 | S | 1 1 0 0 1 1 | 3 |
| 0 1 0 1 0 0 | T | 1 1 0 1 0 0 | 4 |
| 0 1 0 1 0 1 | U | 1 1 0 1 0 1 | 5 |
| 0 1 0 1 1 0 | V | 1 1 0 1 1 0 | 6 |
| 0 1 0 1 1 1 | W | 1 1 0 1 1 1 | 7 |
| 0 1 1 0 0 0 | X | 1 1 1 0 0 0 | 8 |
| 0 1 1 0 0 1 | Y | 1 1 1 0 0 1 | 9 |
| 0 1 1 0 1 0 | Z | 1 1 1 0 1 0 | : |
| 0 1 1 0 1 1 | [ | 1 1 1 0 1 1 | ; |
| 0 1 1 1 0 0 | \ | 1 1 1 1 0 0 | < |
| 0 1 1 1 0 0 | ] | 1 1 1 1 0 1 | = |
| 0 1 1 1 0 1 | ^ | 1 1 1 1 1 0 | > |
| 0 1 1 1 1 1 | — | 1 1 1 1 1 1 | ? |

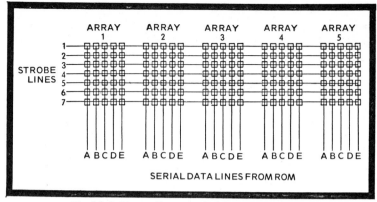

Fig. 12-9. Vertically strobed alphanumeric display.

requires a separate $5 \times 7$ array. Under these conditions, it becomes feasible to drive the arrays by a method known as *scanning*, or *strobing*. This method consists of applying a pulse to one row of all the displays, and simultaneously applying data bits to the individual columns. Each row is sequentially strobed (pulsed). If the strobing occurs at a fairly high rate, the displays do not appear to flicker.

A vertical strobing arrangement for five characters is shown in Fig. 12-9. With the strobing technique, the data bits from the read-only memories must change each time the strobe changes. In this way, the input to a given column appears as a 7-bit serial data stream, which is continuously recirculated. Because the data stream changes in synchronism with the strobe lines, each row of the array always sees the same bit combination.

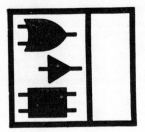

# Appendix I
# Glossary

ABORT—The condition in the computer that results in the skipping of the next sequential instruction.

ACCESS TIME—The time it takes a computer to locate data or an instruction word in its storage section and transfer it to its arithmetic unit where the required computations are performed. The time it takes to transfer information which has been operated on from the arithmetic unit to the location in storage where the information is to be stored. Synonymous with read time.

ACKNOWLEDGE—Indication of the status of data on the input/output lines. Abbreviated as ACK.

ACCUMULATOR—The register and associated equipment in the arithmetic unit of the computer in which arithmetical and logical operations are performed.

ADDRESS—A coded number that specifically designates a computer register or other internal storage location. Information is referenced by its address. Portions of computer control are responsible for directing information to or from an addressed location.

ADDRESSABLE—Capable of being referenced by an instruction.

ADDRESS, DIRECT—An address which indicates the location where the referenced operand is to be found or stored with no reference to an index register or B-box. Synonymous with *first level address*.

ADDRESS, EFFECTIVE—(1) A modified address. (2) The address actually considered to be used in a particular execution of a computer instruction.

ADDRESS INDEXED—An address that is to be modified or has been modified by an index register or similar device. Synonymous with *variable address*.

ADDRESS, ONE-PLUS-ONE—An instruction system having the property that each complete instruction includes an operation and two addresses, one for the location of a register in the storage containing the item to be operated upon, and one for the location containing the next instruction.

ADDRESS, RELATIVE—(1) An address to which the base address must be added in order to find the machine address. (2) An address used for convenience when constructing a program but not the actual address in the final program.

ADP—Automatic Data Processing.

ALGEBRA, BOOLEAN—A process of reasoning, or a deductive system of theorems using a symbolic logic, and dealing with classes, propositions, or on-off circuit elements. It employs symbols to represent operators such as AND, OR, NOT, etc., to permit mathematical calculation. Named after George Boole, famous English mathematician (1815-1864).

ALPHANUMERIC—A contraction of alphabetic-numeric.

ANALOG—The representation of numerical quantities by means of physical variables; e.g., translation, rotation, voltage, or resistance. Contrasted with *digital*.

ANALOG COMPUTER—See *computer, analog*.

ANALYST—A person skilled in the definition of and the development of techniques for the solving of a problem; especially those techniques for solutions on a computer.

AND—Same as *operator, AND*.

AND CIRCUIT—Same as *gate, AND*.

AND GATE—See *gate, AND*.

**AND OPERATOR**—See *operator, AND*.

**ARITHMETIC**—A section within the computer where reasonable processes such as addition, subtraction, multiplication, and division are performed, and operands and results are stored temporarily.

**ARITHMETIC, FIXED POINT**—(1) A method of calculation in which operations take place in invariant manner, and in which the computer does not consider the location of the radix point. This is illustrated by desk calculators or slide rules, with which the operator must keep track of the decimal point. Similarly with many automatic computers, in which the location of the radix point is the programmer's responsibility. Contrasted with *arithmetic, floating point*. (2) A type of arithmetic in which the operands and results of all arithmetic operations must be properly scaled so as to have a magnitude between certain fixed values.

**ARITHMETIC, FLOATING POINT**—A method of calculation which automatically accounts for the location of the radix point. This is usually accomplished by handling the number as a signed mantissa times the radix raised to an integral exponent; e.g., the decimal number $+88.3$ might be written as $+.883 \times 10_2$; the binary number $-.0011$ as $-.11 \times 2_{-2}$. Synonymous with *floating decimal arithmetic* and contrasted with *arithmetic, fixed point*.

**ASSEMBLER**—A computer program which operates on symbolic input data to produce from such data machine instructions by carrying out such functions as : translation of symbolic operation codes into computer operating instructions; assigning locations in storage for successive instructions; or computation of absolute addresses from symbolic addresses. An assembler generally translates input symbolic codes into machine instructions item for item, and produces as output the same number of instructions or constants which were defined in the input symbolic codes. Synonymous with *assembly routine*; *assembly program* and related to *compiler*.

**ASYNCHRONOUS**—Pertaining to a lack of time coincidence in a set of repeated events.

**AUTOMATIC**—(1) The implementation of processes by automatic means; (2) the theory, art, or technique of making a process more automatic; (3) the investigation, design, development, and application of methods of rendering processes automatic, self-moving, or self-controlling.

**BASE**—Same as (radix).

**BINARY**—A characteristic, property, or condition in which there are but two possible alternatives; e.g., the binary number system using 2 as its base and using only the digits zero (0) and one (1).

**BINARY NUMBER SYSTEM**—A number system with two symbols ("0" and "1") that has two as its base just as the decimal system uses ten symbols ("0, 1, ...,9") and a base of ten. (See also *positional notation* and *radix*.)

**BIONICS**—The application of knowledge gained from the analysis of living systems to the creation of hardware that will perform functions in a manner analogous to the more sophisticated functions of the living system.

**BI-STABLE**—The capability of assuming either of two stable states, hence, of storing one bit of information.

**BIT**—A binary digit, zero or one, represented in the computer by the condition (set or clear) of a stage.

**BIT RATE**—See (rate, bit).

**BOOKKEEPING OPERATION**—A computer operation which does not directly contribute to the result; i.e., arithmetical, logical, and transfer operations used in modifying the address section of other instructions, in counting cycles and in rearranging data. Synonymous with *red tape operation*.

**BOOLEAN ALGEBRA**—See *algebra, Boolean*.

**BOOTSTRAP**—A technique for loading the first few instructions of a routine into storage; then using these instructions to bring in the rest of the routine. This usually involves either the entering of a few instructions manually or the use of a special key on the console.

BORROW—A borrow in subtraction is the additional subtraction of a one from the next partial difference and is initiated when a digit of the minuend is zero and the corresponding digit of the subtrahend is one. In a binary system of modules $2k-1$, where k is the number of stages in a register, the borrow produced from the left-most digit $2k-1$ of the minuend is called the end-around borrow. A final correction is made by applying the end-around borrow to the partial difference of the rightmost digits.

BRANCH POINT—A point in a program or instruction where a decision is made on the basis of arithmetic results. The results of the decision indicates whether the main program is to be continued or branched to a different program. See also (jump).

B-REGISTER—(1) Same as *index register*.

BUFFER—Mode of operation that involves interregister or interequipment data trasfer, usually connects an input and output device with the main or high-speed storage.

BYTE—(1) A generic term to indicate a measurable portion of consecutive binary digits; e.g., an 8-bit or 6-bit byte. (2) A group of binary digits usually operated upon as a unit.

CAPACITY—The upper and lower limits of the numbers that may be processed in a computer register.

CAPACITY, STORAGE—The number of elementary pieces of data that can be contained in a storage device. Frequently defined in terms of characters in a particular code or words of a fixed size that can be so contained. Synonymous with *memory capacity*.

CELL—The storage for one unit of information, usually one character or one word.

CHANNEL—(1) A path along which information, particularly a series of digits or characters, may flow. (2) A path for electrical communication. (3) A band of frequencies used for communication.

CHARACTER—(1) One symbol of a set of elementary symbols such as those corresponding to the keys on a typewriter. The symbols usually include the decimal digits 0 through 9, the letters A through Z, punctuation marks, operation symbols, and any other single symbols which a computer may read, stores, or write. (2) The electrical, magnetic, or mechanical profile used to represent a character in a computer, and its various storage and peripheral devices. A character may be represented by a group of other elementary marks, such as bits or pulses.

CLEAR (VERB)—To restore a storage or memory device to the zero state.

CLOCK—(1) A master timing device used to provide the basic sequencing pulses for the operation of a synchronous computer; (2) a register which automatically records the progress of real time, or perhaps some approximation to it, records the number of operations performed, and whose contents are available to a computer program.

CLOCK RATE—See *rate, clock*.

COBOL—Common Business Oriented Language.

CODE—(1) A system of symbols for meaningful communication. (2) A system of symbols for representing data or instructions in a computer or a tabulating machine. (3) To translate the program for the solution of a problem on a given computer into a sequence of machine language or psuedo instructions and addresses acceptable to that computer. (4) A machine language program.

CODED PROGRAM—A procedure for solving a problem by means of a digital computer. The program may vary in detail from a mere outline of the procedure to an explicit list of instructions coded in machine language. See also (program).

COINCIDENCE GATE—A circuit with the ability to produce an output which is dependent upon a specified type of or the coincident nature of the input; e.g., an AND gate has an output pulse when there are pulses in time coincidence at all inputs; an OR gate has an output pulse when there are pulses in time coincidence at all inputs; an OR gate has an output when any one or any combination of input pulses occur in time coincidence. Any gate may contain a number of inhibits, in which there is no output under any condition of input if there is time coincidence of an inhibit or except signal.

COLLATOR—A device used to collate or merge sets or decks of cards or other units into a sequence.

COLUMN—In positional notation, a position corresponding to a given power of the radix. A digit located in any particular column is a coefficient of a corresponding power of the radix.

COMMAND—One of a set of signals or groups of signals resulting from an instruction. Commands initiate the individual steps of the instruction. See also *instruction*.

COMPARATOR—(1) A device for comparing two different transcriptions of the same information to verify the accuracy of transcription, storage, arithmetic operation or other processes, in which a signal is given dependent upon some relation between two items, i.e., one item is larger than, smaller than, or equal to the other. (2) A form of verifier.

COMPILE—To produce a machine language routine from a routine written in source language by selecting appropriate subroutines from a subroutine library, as directed by the instructions or other symbols of the original routine, supplying the linkage which combines the subroutines into a workable routine and translating the subroutines and linkage into machine language. The complied routine is then ready to be loaded into storage and run; i.e., the compiler does not usually run the routine it produces.

COMPILER—A computer program more powerful than an assembler. In addition to its translating function which is generally the same process as that used in an assembler it is able to replace certain items of input with series of instructions, usually called subroutines. Thus, where an assembler translates item for item, and produces as output the same number of instructions or constants which were put into it, a complier will do more than this. The program which results from compiling is a translated and expanded version of the original. Synonymous with *compiling routine* and related to *assember*.

COMPLEMENT—(1) A quantity expressed to the base N, which is derived from a given quantity by a particular rule; frequently used to represent the negative of the given quantity. (2) A complement on N, obtained by subtracting each digit of the given quantity from N-1, adding unity to the least significant digit, and performing all resultant carrys; e.g., the two's complement of binary 11010 is 00110; the tens complement of decimal 456 is 544. (3) A complement on N-1, obtained by subtracting each digit of the given quantity from N-1; e.g., the one's complement of binary 11010 is 00101; the nines complement of decimal 456 is 543.

COMPUTER—A device capable of accepting information, applying prescribed processes to the information, and supplying the results of these processes. It usually consists of input and output devices, storage, arithmetic, and logical units, and a control unit.

COMPUTER, ANALOG—A computer which represents variables by physical analogies. Thus, any computer which solves problems by translating physical conditions such as flow, temperature, pressure, angular position, or voltage into related mechanical or electrical quantities and uses mechanical or electrical equivalent circuits as an analog for the phsycial phenomenon being investigated. In general it is a computer which uses an analog for each variable and produces analogs as output. Thus, an analog computer measures continuously whereas a digital computer counts discretely.

COMPUTER ASYNCHRONOUS—A computer in which the performance of each operation starts as a result of a signal either that the previous operation has been completed, or that the parts of the computer required for the next operation are now available. Contrasted with *computer synchronous*.

COMPUTER, DIGITAL—A computer which processes information represented by combinations of discrete or discontinuous data as compared with an analog computer for continuous data. More specifically, it is a device for performing sequences of arithmetic and logical operations, not only on data, but its own program. Still more specifically, it is a stored program digital computer capable of performing sequences of internally stored instructions, as opposed to calculators, such as card programmed calculators on which the sequence is impressed manually.

COMPUTER, FIXED PROGRAM—A computer in which the sequence of instructions are permanently stored or wired in, and perform automatically and are not subject to change either by the computer or the programmer, except by rewiring or changing the storage input. Synonymous with *special purpose computer*.

COMPUTER, GENERAL PURPOSE—A computer designed to solve a large variety of problems; e.g., a stored program computer which may be adapted to any of a very large class of applications.

COMPUTER, SYNCHRONOUS—A computer in which all operations and events are controlled by equally spaced pulses from a clock. Contrasted with *computer, asynchronous*.

COMPUTER, WIRED PROGRAM—A computer in which the instructions that specify the operations to be performed are specified by the placement and interconnection of wires. The wires are usually held by a removable control panel, allowing flexibility of operation, but the term is also applied to permanently wired machines which are then called fixed program computers.

CONDITIONAL TRANSFER—See *transfer, conditional*.

CONTROL—The computer circuits that affect the carrying out of instructions in the proper sequence, the interpretation of each instruction, and the application of the proper commands to other sections and circuits in accordance with the interpretation.

CONTROL WORD—See *word, control*.

CONVERTER—A device which converts the representation of information, or which permits the changing of the method for data processing from one form to another; e.g., a unit which accepts information from punch cards and records the information on magnetic tape, and possibly including editing facilities.

COUNTER—A device capable of increasing or decreasing its own contents upon receipt of separate input signals.

CORE MATRIX—An array of cores, each of which represents the same column for each storage register in the magnetic core storage system.

CORE STORAGE—A type of storage system in which the magnetic core is the basic memory element.

CPU—Central Processing Unit.

CRYOGENICS—The field of technology in which devices utilizing properties assumed by metals at absolute zero are used. At these temperatuares large current changes can be obtained by relatively small magnetic field changes.

CYBERNETICS—The field of technology involved in the comparative study of the control and intracommunication of information handling machines and nervous systems of animals and man in order to understand and improve communication.

DATA—A general term used to denote any or all facts, numbers, letters, and symbols, or facts that refer to or describe an object, idea, condition, situation, or other factors. It connects basic elements of information which can be processed or produced by a computer. Sometimes data is considered to be expressible only in numerical form, but information is not so limited. Related to *information*.

DATA, RAW—Data which has not been processed. Such data may or may not be in machine-sensible form.

DEBUG—To isolate and remove all malfunctions from the computer, or all mistakes from a routine or program.

DECIMAL, BINARY CODED—Describing a decimal notation in which the individual decimal digits are represented by a pattern of ones and zeros; e.g., in the 8-4-2-1 coded decimal notation, the number twelve is represented as 0001 0010 for 1 and 2, respectively, whereas in pure or straight binary notation is represented as 1100. Related to *binary*.

**DECODER**—(1) A device which determines the meaning of a set of signals and initiates a computer operation based thereon. (2) A matrix of switching elements which selects one or more output channels according to the combination of input signals present. Contrasted with *encoder*.

**DECREMENT**—The quantity by which a variable is decreased.

**DENSITY CHARACTER**—The number of characters that can be stored per unit of length.

**DIAGNOSTIC ROUTINE**—See *routine, diagnostic*.

**DIAGRAM, FLOW**—Same as *flow chart*.

**DIAGRAM, VENN**—A diagram in which each point represents an individual. Sets are represented by closed regions including all members of the set and excluding all nonmembers. The diagram is used to facilitate determination whether several sets include or exclude the same individuals.

**DIGIT**—One of a set of characters used as coefficients or powers of the radix in the positional notation of numbers.

**DIGITAL COMPUTER**—See *computer, digital*.

**DISC, MAGNETIC**—A storage device on which information is recorded on the magnetizable surface of a rotating disc. A magnetic disc storage system is an array of such devices, with associated reading and writing heads which are mounted on movable arms. Related to *storage, disc*.

**DRUM, MAGNETIC**—A cylinder having a surface coating of magnetic material, which stores binary information by the orientation of magnetic dipoles near or on its surface. Since the drum is rotated at a unifrom rate, the information stored is available periodically as a given portion of the surface moves past one or more flux detecting devices called heads located near the surface of the drum.

**DRUM**—Transfer of information from one piece of equipment to another (normally from computer to external equipment such as paper tape, high-speed printer, etc.)

**DYNAMIC STORAGE**—The storage of data on a device or in a manner that permits the data to move or vary with time, and thus the data is not always available instantly for recovery; e.g., acoustic delay line, magnetic drum, or circulating or recirculating of information in a medium. Synonymous with *dynamic memory*

**EDP**—Electronic Data Processing.

**EFFECTIVE ADDRESS**—See *address, effective*.

**ELECTROSTATIC STORAGE**—(1) The storage of data on a dielectric surface such as the screen of a cathode ray tube, in the form of the present or absence of spots bearing electrostatic charges, that can persist for a short time after the electrostatic charging force is removed. (2) A storage device so used.

**ENABLE**—The application of a pulse that prepares a circuit from some subsequent action.

**ENCODER**—A device capable of translating from one method of expression to another method of expression, e.g., translating a message, "add the contents of A to the contents of B," into a series of binary digits. Contrasted with *decoder*.

**EOF**—End of File.

**ERASE**—To replace all the binary digits in a storage device by binary zeroes.

**EXCLUSIVE OR OPERATOR**—A logical operator which has the property that, if P and Q are two statements, then the statement PQ, where the    is the Exclusive OR operator, is true if either P or Q, but not both are true, and false if P and Q are both false or both true.

**EXECUTION TIME**—The portion of an instruction cycle during which the actual work is performed or operation executed; i.e., the time required to decode and perform an instruction. Synonymous with *time, instruction (2)*.

**FAULT**—(1) The condition resulting from the execution of an improper instruction, (2) a malfunction.

**FILE**—An organized collection of information directed toward some purpose. The records in a file may or may not be sequenced according to a key contained in each record.

**FIXED POINT ARITHMETIC**—See *computer, fixed program*.

**FIXED PROGRAM COMPUTER**—See *computer, fixed program*.

**FIXED WORD LENGTH**—Having the property that a machine word always contains the same number of characters or digits.

**FLIP-FLOP**—(1) A bistable device; i.e., a device capable of assuming two stable states. (2) A bistable device which may assume a given stable state deznding upon the pulse of history of one or more input points and having one or more output points. The device is capable of storing a bit of information. (3) A control device for opening or closing gates; i.e., a toggle. Synonymous with *Eccles-Jordon circuit* and *Eccles-Jordan trigger*.

**FLOATING POINT ARITHMETIC**—See *arithmetic, floating point*.

**FLOW CHART**—A graphic representation of the major steps of work in process. The ilustrative symbols may represent documents, machines, or actions taken during the process. The area of concentration is on where or who does what rather than how it is to be done. Synonymous with *process chart* and *flow diagrm*.

**FLOW DIAGRAM**—Same as *flow chart*.

**FORMAT**—The predetermined arrangement of characters, fields, lines, page numbers, and punctuation marks, usually on a single sheet or in a file. This refers to input, output, and fils.

**FORTRAN**—A programing language designed for problems which can be expressed in algebraic notation, allowing for exponentiation and up to three subscripts. The FORTRAN compiler is a routine for a given machine which accepts a program written in FORTRAN source language and produces a machine language routine object program. FORTRAN II added considerably to the power of the original language by giving it the ability to define and use almot unlimited hierarchies of subroutines, all sharing a common storage region, if desired. Later improvements have added the ability to use Boolean expressions, and some capabilities for inserting symbolic machine language sequences within a source program.

**FUNCTION CODE**—The portion of the instruction word that specifies to the control section the particular instruction which is to be performed.

**GATE, AND**—A signal circuit with two or more input wires in which the output wire gives a signal, if and only if, all input wires receive coincident signals. Synonymous with *and circuit* and clarified by *conjunction*.

**GATE, OR**—An electrical gate or mechanical device which implements the logical OR operator. An output signal occurs whenever there is one or more inputs on a multi-channel input. An OR gate performs the function of the logical "inclusive OR Operator." Synonymous pth *OR circuit* and clarified by *disjunction*.

**GENERAL PURPOSE COMPUTER**—See *computer, general purpose*.

**GRAY CODE**—A binary code in which sequential numbers are represented by expressions which are the same xcept in one place, and in that place differ by one unit.

**HALF-ADDER**—A circuit having two output points, S and C, representing sum and carry, and two input points A and B, representing addend and augend, such that the output is related to the input according to the following table:

| INPUT | | OUTPUT | |
|---|---|---|---|
| A | B | S | C |
| 0 | 0 | 0 | 0 |
| 0 | 1 | 1 | 0 |
| 1 | 0 | 1 | 0 |
| 1 | 1 | 0 | 1 |

A and B are arbitrary input pulses, and S and C are sum without carry and carry, respectively. Two half-adders properly connected, may be used for performing binary addition and form a full serial adder.

**HALF-SUBTRACT**—The bit-by-bit subtraction of two binary numbers with no regard for borrows. Abbreiated as HS. The complement of the half-subtract is "half-subtract-not," abbreviated as HSN.

INCLUSIVE-OR OPERATOR—A logical operator which has the property that P or Q is true if P or Q or both is true; wyn the term OR is used alone, as in OR-gate, the inclusive-OR is usually implied.

INDEX REGISTER—A register which contains a quantity which may be used to modify addresses. Synonymous with *B-register*, and *B-box*.

INSTRUCTION WORD—Designators specifying a particular function. The designators are listed below as they appear from highest to lowest order significance in the instruction word of a representative computer:

      f=Function Code
      j=Branch Condition
      k=Operand Interpretation
      b=Address Modification
      y=Operand Address

INPUT/OUTPUT—A section providing the means of communication between the computer and external equipment or other computers. Input and output operations involve units of external equipment, certain registers in the computer, and portions of the computer control section. Abbreviated I/O.

INTERFACE—A common boundary between automatic data processing systems or parts of a single system.

INTERRUPT—(1) Internal: indicates the termination of an input or output buffer; (2) External: signal on the data lines that requires computer attention.

I/O—The abbreviation for input/output. Synonymous with *input/output*.

JUMP—An instruction that specifies the location of the next instruction and directs the computer. A jump is used to alter the normal sequence control of the computer. Under certain conditions, a jump may be contingent upon manual intervention.

JUMP, CONDITIONAL—An instruction which, if a specified condition or set of conditions is satisfied, is interpreted as an unconditional transfer. If the condition is not satisfied, the instruction causes the computer to proceed in its normal sequence of control. A conditional transfer also includes the testing of the condition. Synonymous with *conditional jump* and *conditional branch*.

JUMP, UNCONDITIONAL—An instruction which switches the sequence of control to some specified location. Synonymous with *unconditional branch*; *unconditional jump* and *unconditional transfer of control*.

KEY—(1) A group of characters which identifies or is part of a record or item, thus, any entry in a record or item can be used as a key for collating or sorting purposes.(2) A marked lever manually operated for copying a character; e.g., a typewriter, paper tape perforator, card punch, manual keyboards, digitizer or manual word generator. (3) A lever or switch on a computer console for the purpose of manually altering computer action.

LIBRARY—(1) A collection of information available to a computer, usually on magnetic tapes; (2) a file of magnetic tapes.

LOAD—To enter information into either the computer or a storage location.

LOGIC—(1) The science dealing with the criteria or formal principles of reasoning and thought. (2) The systematic scheme which defines the interactions of signals in the design of an autowtic data processing system. (3) The basic principles and application of truth tables and interconnection between logical elements required for arithmetic computation in an automatic data processing system. Related to ymbolic logic.

LOGICAL SUM—The bit-by-bit addition of two binary numbers with no regard for carries. Abbreviated as LS. The complement of the logical sum is "logical sum not," abbreviated as LS.

LOOP—(1) Self-contained series of instructions in which the last instruction can modify and repeat itself until a terminal condition is reached. The productive instructions in the loop generally manipulate the operands, while bookkeeping instructions modify the productive instructions, and keep count of the number of repetitions. A loop may contain any number of conditions for termination. The equivalent of a loop can be achieved by the technique of straight line codin. whereby the repetition of productive and bookkeeping operations is accomplished by explicitly writing the instruction for each repetition.

MALFUNCTION—Nonoperation of the computer due to component failure.

MARGIN—A measure of the tolerance of a circuit; he range between an established operating point and the point at which the circuit first starts to fail.

MASK—A machine word that specifies which parts of another machine word are to be operated upon, thus, the criterion for an external command. Synonymous with (extractor).

MASKING—(1) The process of extracting a non-word group or a field of characters from a word or a string of words, (2) the process of setting internal program controls to prevent transfers which otherwise would occur upon setting of internal machine latches.

MASTER CLOCK—The primary source of timing signals.

MEMORY—Same as (storage).

MNEMONIC—Pertaining to the assisting, or intending to assist, human memory; thus, a mnemonic term, usually an abbreviation, that is easy to remember; e.g., mpy for multiply and acc for accumulator.

MODE—(1) A computer system of data representation; e.g., the binarymode. (2) A selected mode of computer operation.

MODEL, MATHEMATICAL—The general characterization of a process, object, or concept, in terms of mathematics, which enables the relatively simple manipulation of variables to be accomplished in order to determine how the process, object, or concept would behave in different situations.

MODIFY—(1) To alter a portion of an instruction so its interpretation and execution will be other than normal. The modification may permanently change the instruction or leave it unchanged and affect only the current execution. The most frequency modification is that of the effective address through use of index registers. (2) To alter a subroutine according to a defined parameter.

MODULUS—The number of permissible numbers used in a process or system. For example, if only the integers from $-15$ to $+15$ inclusive are considered, 31 is the modulus of this set of numbers.

MOST SIGNIFICANT DIGIT—The first digit from the left, different from zero.

NONDESTRUCTIVE READ—See read, nondestructive.

NONVOLATILE STORAGE—Storage media that retains information kring the absence of power. These media include magnetic tapes, drums, and magnetic cores.

NOTATION—(1) The act, process, or method of representing facts or quantities by a system or set of marks, signs, figures, or characters. (2) A system of such symbols or abbreviations used to express technical facts or quantities; as mathematical notation. (3) An annotation; note.

OBJECT LANGUAGE—A language which is the output of an automatic coding routine. Usually object language and machine language are the same; however, a series of steps in an automatic coding system may involve the object language of one step serving as a source language for the next step and so forth.

OCTAL NUMBER—Numbers in a system using eight symbols, 0, 1, ..., 6, 7, with eight as it's base.

OFF-LINE—Descriptive of a system and of the peripheral equipment or devices in a system in which the operation of peripheral equipment is not under the control of the central processing unit.

ON-LINE—Descriptive of a system and of the peripheral equipment or devices in a system in which the operation of such equipment is under control of the central processing unit, and in which information reflecting current activity is introduced into the data processing system as soon as it occurs. Thus, directly in-line with the main flow of transavion processing.

OPEN-ENDED—The quality by which the addition of new terms, subject headings, or classifications does not disturb the pre-existing system.

OPERAND—A quantity entering or arising in an instruction. An operand may be an argument, a result, a parameter, or an indication of the location of the next instruction, as opposed to the operation code or symbol itself. It may even be the address portion of an instruction.

OPERATION CODE—The part of a computer instruction word which specifies, in coded form, the operation to be performed.

OPERATOR, AND—(1) A logical operator which has the property that if P is a statement and Q is a statement, the P AND Q is true if both statements are true, false if either is false or both are false, Truth is normally expressed by the value 1, falsity by 0. The AND operator is often represented by a centered dot (P . Q), by no sign (PQ), by an inverted "u" or logical product symbol (PnQ), or by the letter "x" or multiplication symbol (P×Q). Note that the letters AND are capitalized to differemiate between the logical operator AND and the word *and* in common usage. (2) The logical operation which makes use of the AND operator or logical product. Synonymous with *and; logical multiply* and clarified by *conjunction*.

OPERATOR, OR—A logical operator which has the property such that if P or Q are two statements, then the statement P OR Q is true or false varies according to the following table of possible combinations:

| P | Q | P or Q |
|---|---|---|
| False | True | True |
| True | False | True |
| True | True | True |
| False | False | False |

OR GATE—See *gate, or*.

OR OPERATOR—See *operator, or*.

OVERFLOW—The condition which arises when the result of an arithmetic operation exceeds the capacity of the storage space alloted in a digital computer.

PACK—To include several short items of information into one machine item or word by utilizing different sets of digits to specify each brief item.

PADDING—A technique used to fill out a block of information with dummy records.

PARALLEL TRANSMISSION—The system of information transfer in whch the characters of a word are transmitted simultaneously over separate lines.

PARAMETER—(1) A quantity in a subroutine, whose value specifies or partly specifies the process to be performed. It may be given different values when the subroutine is used in different main routines or in different parts of one main routine, but which usually remains unchanged throughout any one such use. Related to *parameter, program*. (2) A quantity used in a generator to specify machine configuration, designate subroutines to be included, or otherwise to describe the desired routine to be generated. (3) A constant or a variable in mathematics, which remains constant during some calculation. (4) A definable characteristic of an item, device, or system.

PARITY BIT—A check bit that indicates whether the total number of binary "1" digits in a character or word (exclusing the parity bit) is odd or even. If a "1" parity bit indicates an odd number of "1" digits, then a "0" bit indicates an even number of them. If the total number of "1" bits, including the parity bit, is always even, the system is called an even parity system. In an odd parity system, the total number of "1" bits, including the parity bit, is always odd.

PARTIAL CARRY—A system of executing the carry process in which the carries that arise as a result of a carry are not allowed to be transmitted to the next higher stage.

PERIPHERAL EQUIPMENT—The auxiliary machines which may be placed under the control of the central computer. Examples of this are card readers, card punches, magnetic tape feeds and high-speed printers. Peripheral equipment may be used on-line or off-line depending upon computer design, job requirements, and economics.

PING-PONG—The programming technique of using two magnetic tape units for multiple reel files and switching automatically between the two until the complete file is processed.

PLOTTER—A visual display or board in which a dependent variable is graphed by an automatically controlled pen or pencil as a function of one or more variables.

POSITIONAL NOTATION—A method for expressing a quantity, using two or more figures, wherein the successive right to left figures are to be interpreted as coefficients of ascending integer powers of the radix.

PRECISION, DOUBLE—The retention of twice as many digits of a quantity as the computer normally handles; e.g., if a computer, whose basic word consists of 10 decimal digits is called upon to handle 20 decimal digit quantities, then double precision arithmetic must be resorted to.

PROBLEM ORIENTED LANGUAGE—(1) A language designed for convenience of program specification in a general problem area rather than for easy conversion to machine instruction code. The components of such a language may bear little resemblance to machine instructions. (2) A machine independent language where one needs only to state the problem, not the how of solution. Related to *generators, program* and contrasted with *language, procedure oriented*.

PROGRAM—(1) The complete plan for the solution of a problem, more specifically the complete sequence of machine instructions and routines necessary to solve a problem. (2) To plan the procedures for solving a problem. This may involve among other things the analysis of the problem, preparation of a flow diagram, preparing details, testing, and developing subroutines, allocation of storage locations, specification of input and output formats, and the incorporation of a computer run into a complete data processing system. Related to *routine*.

PROGRAM, OBJECT—The program which is the output of an automatic coding system. Often the object program is a machine language program ready for execution, but it may well be in an intermediate language. Synonymous with *target program; object routine* and contrasted with *program source*.

PROGRAM, SOURCE—A computer program written in a language designed for ease of expression of a class of problems or procedures, by humans; e.g., symbolic or algebraic. A generator, assembler translator or compiler routine is used to perform the mechanics of translating the source program into an object program in machine language.

PROGRAMER—A person who prepares problem solving procedures and flow charts and who may also write and debug routines.

PROGRAMING, AUTOMATIC—The method or technique whereby the computer itself is used to transform or translate programing from a language or form that is easy for a human being to produce into a language that is efficient for the computer to carry out. Examples of automatic programing are compiling, assembling, and interpretive routines.

PROGRAMING, INTERPRETIVE—The writing of programs in a pseudo machine language, which is precisely converted by the computer into actual machine language instructions before being performed by the computer.

PROGRAMING, RANDOM ACCESS—Programing without regard to the time required for access to the storage positions called for in the program. Contrasted with *programing, minimum access*.

RADIX—The quantity of characters for use in each of the digital positions of a numbering system. In the more common numbering systems the characters are some or all of the Arabic numerals as follows:

| System Name | Characters | Radix |
|---|---|---|
| BINARY | (0, 1) | 2 |
| OCTAL | (0, 1, 2, 3, 4, 5, 6, 7) | 8 |
| DECIMAL | (0, 1, 2, 3, 4, 5, 6, 7, 8, 9) | 10 |

Unless otherwise indicated, the radix of any number is assumed to be 10. For positive identification of a radix 10 number, the radix to the expressed number; i.e., $126_{(10)}$. The radix of any nondecimal number is expressed in similar fashion; e.g., $11_{(2)}$ and $5_{(8)}$. Synonymous with *base; base number* and *radix number*.

RADIX MINUS 1 COMPLEMENT—Same as *complement (3)*.

RATE, BIT—The rate at which binary digits, or pulses representing them pass a given point on a communications line or channel.

RATE, CLOCK—The time rate at which pulses are emitted from the clock. The clock rate determines the rate at which logical or arithmetic gating is performed with a synchronous computer.

RATE, SAMPLING—The rate at which measurements of physical quantities are made; e.g., if it is desired to calculate the velocity of a missile and its position is measured each millisecond, then the sampling rate is 1000 measurements per second.

RAW DATA—See *data, raw*.

READ-IN—To sense information contained in some source and transmit this information to an internal storage.

READ, NONDESTRUCTIVE—A reading of the information in a register without changing that information.

READOUT—To sense information contained in some internal storage and transmit this informution to a storage external to the computer.

READOUT—Information displayed visually.

READ TIME—Same as *access time*.

REAL TIME—Computer operation with regard to a related process so that the computer results are available to conduct or guide the process.

RECORD—(1) A group of related facts or fields of information treated as a unit, thus a listing of information, usually in printed or printable form; (2) to put data into a storage device.

REGISTER—A hardware device used to store a certain amount of bits or characters. A register is usually constructed of elements such as transistors and usually contains approximately one word of information. Common programming usage demands that a register have the ability to operate upon information and not merely store information; hardware usage does not make the distinction.

REGISTER B—See *index register*.

REGISTER, STORAGE—A register in the storage of the computer, in contrast with a register in one of the other units of the computer.

RELATIVE ADDRESS—See *address, relative*.

RELIABILITY—(1) A measure of the ability to function without failure; (2) the amount of credence placed in a result.

REPERTOIRE OF INSTRUCTIONS—(1) The set of instructions which a computing or data processing system is capable of performing, (2) the set of instructions which an automatic coding system assemblies.

RETURN—The mechanism providing for a return in the usual sense. In particular a set of instructions at the end of a subroutine which permit control to return to the proper point in the main routine.

ROUTINE—A set of coded instructions arranged in proper sequence to direct the computer to perform a desired operation or sequence of operations. A subdivision oof a program consisting of two or more instructions that are functionally related; therefore, a program.

ROUTINE, DIAGNOSTIC—A routine used to locate malfunction in a computer, or to aid in locating mistakes in a computer program. Thus, in general any routine specifically designed to aid in debugging or troubleshooting. Synonymous with *malfunction routine*.

ROUTINE, EXECUTIVE—A routine which controls loading and relocation of routines and in some cases makes use of instructions which are unknown to the general programer. Effectively, an executive routine is part of the machine itself.

ROUTINE, SERVICE—A broad class of routines which are standardized at a particular installation for the purpose of assisting in maintenance and operation of the computer as well as the preparation of programs as opposed to routines for the actual solution of production problems. This class includes monitoring or supervisory routines, assemblers, compilers, diagnostics for computer malfunctions, simulation of peripheral equipment, general diagnostics and input data. The distinguishing quality of service routines is that they are generally standardized so as to meet the servicing needs at a particular installation, independent of any specific production type routine requiring such services.

SAMPLING RATE—See *rate, sampling*.

SCALE—A range of values frequently dictated by the computer word-length or routine at hand.

SEMICONDUCTOR—A solid with an electrical conductivity that lies between the high conductivity of metals and the low conductivity of insulators. Semiconductor circuit elements include crystal diodes and transistors.

SENSE—(1) To examine, particularly relative to a criterion; (2) to determine the present arrangement of some element of hardware, especially a manually-set switch; (3) to read punched holes or other marks.

SENSITIVITY—The degree of response of an instrument or control unit to a change in the incoming signal.

SERIAL—(1) The handling of one after the other in a single facility, such as transfer or store in a digit-by-digit time sequence, or to process a sequence of instructions one at a time, i.e., sequentially. (2) The time sequence transmission of storage, or logical operations on the parts of a word with the same facilities for successive parts.

SERIAL-PARALLEL—(1) A combination of serial and parallel; e.g., serial by character, parallel by bits comprising the character. (2) Descriptive of a device which converts a serial input into a parallel output.

SERVOMECHANISM—A device to monitor an operation as it proceeds, and make necessary adjustments to keep the operation under control.

SET—(1) To place a storage device in a prescribed state. (2) To place a binary cell in the one state. (3) A collection of elements having some feature in common or which bear a certain relation to one another; e.g., all even numbers, geometrical figures, terms in a series, a group of irrational numbers, all positive even integers less than 100 may be a set or a sub-set.

SHIFT—To move the characters of a unit of information columnwise right or left. For a number, this is equivalent of multiplying or dividing by a power of the base of notation.

SHIFT, ARITHMETIC—To multiply or divide a quantity by a power of the number base; e.g., if binary 1101, which represents decimal 13, is arithmetically shifted twice to the left, the result is 110100, which represents 52, which is also obtained by multiplying 13 by 2 twice; on the other hand, if the decimal 13 were to be shifted to the left twice, the result would be the same as multiplying by 10 twice, or 1300.

SHIFT REGISTER—A register in which the characters may be shifted one or more positions to the right or left.

SIGN DIGIT—A binary digit used as a sign.

SKIP INSTRUCTION—An instruction having no effect other than directing the processor to proceed to another instruction designated in the storage portion.

SKIP, TAPE—A machine instruction to space forward and erase a portion of tape when a defect on the tape surface causes a write error to persist.

SOFTWARE—The totality of programs and routines used to extend the capabilities of computers, such as compilers, assemblers, narrators, routines, and subroutines. Contrasted with *hardware*.

SOLID STATE—The electronic components that convey or control electrons within solid materials.

SORTER—A machine which puts items of information into a particular order; e.g., it will determine whether A is greater than, equal to, or less than B and sort or order accordingly.

STORAGE—(1) The term preferred to memory. (2) Pertaining to a device in which data can be stored and from which it can be obtained at a later time. The means of storing data may be chemical, electrical, or mechanical. (3) A device consisting of electronic, electrostatic, electrical, hardware, or other elements into which data may be entered, and from which data may be obtained as desired.

STORAGE, AUXILIARY—A storage device in addition to the main storage of a computer; e.g., magnetic tape, disc, or magnetic drum. Auxiliary storage usually holds much larger amounts of information than the main storage, and the information is accessible less rapidly.

STORAGE, BUFFER—(1) A synchronizing element between two different forms of storage, usually between internal and external. (2) An input device in which information is assembled from external or secondary storage and stored ready for transfer to internal storage. (3) An output device into which information is copied from internal storage. Computation continues while transfers between buffer storage and internal storage or vice versa take place. (4) Any device which stores information temporarily during data transfers.

STORAGE CAPACITY—See *capacity, storage*.

STORAGE DISC—The storage of data on the surface of magnetic discs.

STORAGE, EXTERNAL—(1) The storage of data on a device which is not an integral part of a computer, but in a form prescribed for use by the computer. (2) A facility or device, not an integral part of a computer, on which data usable by a computer is stored such as, off-line magnetic tape units or punch card devices.

STORAGE, INTERNAL—(1) The storage of data on a device which is an integral part of a computer. (2) The storage facilities forming an integral physical part of the computer and directly controlled by the computer. In such facilities all data are automatically accessible to the computer; e.g., magnetic core and magnetic tape on-line.

STORAGE, MAGNETIC—A device or devices which utilize the magnetic properties of materials to store information.

STORAGE, MAGNETIC CORE—A storage device in which binary data is represented by the direction of magnetization in each unit of an array of magnetic material, usually in the shape of torodial rings, but also in other forms such as thin film. Synonymous with *core storage*.

STORAGE, MAGNETIC DRUM—The storage of data on the surface of magnetic drums.

STORAGE, MAGNETIC TAPE—A storage device in which data is stored in the form of magnetic spots on metal or coated plastic tape. Binary data are stored as small magnetized spots arranged in column form across the width or the tape. A read-write head is usually associated with each row of magnetized spots so that one column can be read or written at a time as the tape traverses the head.

STORAGE, MAIN—Usually internal, and the fastest storage device of a computer. The one from which instructions are executed. Contrasted with *storage, auxiliary*.

STORAGE, NONVOLATILE—A storage medium, which retains information in the absence of power and which may be made available upon restoration of power, e.g., magnetic tapes, cores, drums, and discs. Contrasted with *storage, volatile*.

STORAGE, PROGRAM—A portion of the internal storage reserved for the storage of programs, routines, and subroutines. In many systems protection devices are used to prevent inadvertent alteration of the contents of the program storage.

STORAGE, RANDOM ACCESS—A storage technique in which the time required to obtain information is independent of the location of the information most recently obtained. This strict definition must be qualified by the observation that we usually mean relatively random.

STORAGE REGISTER—See *register, storage*.

STORAGE, SECONDARY—The storage facilities not an integral part of the computer but directly connected to and controlled by the computer; e.g., magnetic drum and magnetic tapes.

STORAGE, SEQUENTIAL ACCESS—A storage technique in which the items of information stored become available only in one after the other sequence, whether or not all the information or only some of it is desired; e.g., magnetic tape storage.

**STORAGE, SERIAL**—A storage technique in which time is one of the factors used to locate any given bit, character, word, or group of words appearing one after the other in time sequence, and in which access time includes a variable latency or waiting time of from zero to many word times. A storage is said to be serial by word when the individual bits comprising a word appear serially in time; or a storage is serial by character when the characters representing coded decimal or other nonbinary numbers appear serially in time; e.g., magnetic drums are usually serial by word but may be serial by bit, or parallel by bit, or serial by character and parallel by bit.

**STORAGE, VOLATILE**—A storage medium in which information cannot be retained without continuous power dissipation. Contrasted with *storage, non-volatile*.

**STORAGE, WILLIAMS TUBE**—A cathode ray tube used as an electrostatic storage device and of the type designed by F. C. Williams, University of Manchester, England. Synonymous with *Williams tube storage*.

**STORE**—(1) To transfer an element of information to a device from which the unaltered information can be obtained at a later time, (2) to retain data in a device from which it can be obtained at a later time.

**SUBPROGRAM**—A part of a larger program which can be converted into machine language independently.

**SUBROUTINE**—(1) The set of instructions necessary to direct the computer to carry out a well defined mathematical or logical operation. (2) A subunit of a routine. A subroutine is often written in relative or symbolic coding even when the routine to which it belongs is not. (3) A portion of a routine that causes a computer to carry out a well-defined mathematical or logical operation. (4) A routine which is arranged so that control may be transferred to it from a master routine and so that, at the conclusion of the subroutine, control reverts to the master routine. Such a subroutine is usually called a closed subroutine. (5) A single routine may simultaneously be both a subroutine with respect to another routine and a master routine with repsect to a third. Usually control is transferred to a single subroutine from more than one place in the master routine and the reason for using the subroutine is to avoid having to repeat the same sequence of instructions in different places in the master routine.

**SUBROUTINE, DYNAMIC**—A subroutine which involves parameters, such as decimal point position or item size, from which a relatively coded subroutine is derived. The computer itself is expected to adjust or generate the subroutine according to the parametric values chosen.

**SWITCH, TOGGLE**—(1) An electronically operated circuit that holds either of two states until changed. (2) A manually operated electric switch, with a small projecting knob or arm that may be placed in either of two positions, "on" or "off," and will remain in that position until changed.

**TABLE**—A collection of data in a form suitable for ready reference, frequently as stored in sequenced machine locations or written in the form of an array of rows and columns for easy entry and in which an intersection of labeled rows and columns serves to locate a specific piece of data or information.

**TABLE, TRUTH**—A representation of a switching function, or truth function, in which every possible configuration of argument values 0, 1, or true-false is listed, and beside each is given the associated function value 0-1 or true-false. The number of configurations is $2n$, where n is the number of arguments.

**TAPE**—A strip of material, which may be punched, coated, or impregnated with magnetic or optically sensitive subtances, and used for data input, storage, or output. The data are stored serially in several channels across the tape transversely to the reading or writing motion.

**TERNARY**—Pertaining to a system of notation utilizing the base of 3.

**TEST, DIAGNOSTIC**—The running of a machine program or routine for the purpose of discovering a failure or a potential failure of a machine element, and to determine its location or its potential location.

**TIME, EXECUTION**—See *execution time*.

**TIME, INSTRUCTION**—(1) The portion of an instruction cycle during which the control unit is analyzing the instruction and setting up to perform the indicated operation; (2) same as *execution time*.

TIME, REAL—See *real time*.

TIME SHARING—The use of a device for two or more purposes during the same overall time interval, accomplished by interspersing component actions in time.

TOGGLE—(1) A fliplop. (2) Pertaining to a manually operated on-off switch; i.e., a two position switch. (3) Pertaining to flip-flop, see-saw, or bistable action.

TRANSFER—(1) The conveyance of control from one mode to another by means of instructions or signals. (2) The conveyance of data from one place to another. (3) An instruction for transfer. (4) To copy, exchange, read, record, store, transmit, transport, or write data. (5) An instruction which provides the ability to break the normal sequential flow of control. Synonymous with *jump*, and *control transfer*.

TRANSFER, BLOCK—The conveyance of a group of consecutive words from one place to another.

TRANSFER, CONDITIONAL—An instruction which, if a specified conditions is satisfied, is interpreted as an unconditional transfer. If the condition is not satisfied, the instruction causes the computer to proceed in its normal sequence of control. A conditional transfer also includes the testing of the condition. Synonymous with *conditional jump* and *conditional branch*.

TRANSFER, PARALLEL—A method of data transfer in which the characters of an element of information are transferred simultaneously over a set of paths.

TRANSFER, UNCONDITIONAL—See *jump, unconditional*.

TRANSFLUXOR—A magnetic core having two or more openings. Control of the magnetic flux in the various legs of the magnetic circuits and the binary magnetic characteristics of the material permits storage.

TRANSIENT—(1) A physical disturbance, intermediate to two steady-state conditions. (2) Pertaining to rapid change. (3) A build-up or breakdown in the intensity of a phenomenon until a steady state condition is reached. The time rate of change of energy is finite and some form of energy storage is usually involved.

TRANSISTOR—An electronic device utilizing semiconductor properties to control the flow of currents.

TRANSLATOR—(1) A program whose input is a sequence of statements in some language and whose output is an equivalent sequence of statements in another language. (2) A translating device.

TRANSMISSION, SERIAL—To move data in sequence, one character at a time as contrasted with parallel transmissions.

TRANSPORT, TAPE—The mechanisum which moves magnetic or paper tape past sensing and recording heads and usually associated with data processing equipment. Synonymous with *tape transport*, *tape drive*, and *feed*, *tape*.

TRUTH TABLE—See *table, truth*.

ULTRASONICS—The field of science devoted to frequencies of sound above the human audio range; i.e., above 20 kHz.

UNCONDITIONAL JUMP—Same as *jump, unconditional*.

UNIT, ARITHMETIC—The portion of the hardware of a computer in which arithmetic and logical operations are performed. The arithmetic unit generally consists of an accumulator, some special registers for the storage of operands, and results supplemented by shifting and sequencing circuitry for implementing multiplication, division, and other desired operations.

UNIT CONTROL—The portion of a computer which directs the sequence of operations, interprets the coded instructions, and initiates the proper commands to the computer circuits preparatory to execution.

UNPACK—To separate various sections of a tape record or computer word, and store them in separate locations. The sections usually correspond to format fields within the record or word.

UPDATE—(1) To put into a master file changes required by current information or transactions, (2) to modify an instruction so that the address numbers it contains are increased by a stated amount each time the instruction is performed.

VARIABLE—(1) A quantity which can assume any of the numbers of some set of numbers, (2) a condition, transaction, or event which changes or may be changed as a result of processing additional data thru the system.

VARIABLE WORD-LENGTH—See *word length, variable*.

VECTOR—A quantity having magnitude and direction, as contrasted with scalar which has quantity only.

VENN DIAGRAM—See *diagram, Venn*.

VERIFIER—A device on which a record can be compared or tested for identity character-by-character with a retranscription or copy as it is being prepared.

VOCABULARY—A list of operating codes or instructions available to the programmer for writing the program for a given problem for a specific computer.

VOLATILE STORAGE—See *storage, volatile*.

WILLIAMS TUBE—See *storage*, Williams *tube*.

WORD—An ordered set of characters which occupies one storage location and is treated by the computer circuits as a unit and transferred as such. Ordinarily a word is treated by the control unit as an instruction, and by the arithmetic unit as a quantity. Word lengths may be fixed or variable depending on the particular computer.

WORD, CONTROL—A word, usually the first or last of a record, or first or last word of a block, which carries indicative information for the following words, records, or blocks.

WORD, DATA—A word which may be primarily regarded as part of the information manipulated by a given program. A data word may be used to modify a program instruction, or to be arithmetically combined with other data words.

WORD LENGTH, FIXED—Same as *fixed word length*.

WORD LENGTH, VARIABLE—Having the property that a machine word may have a variable number of characters. It may be applied either to a single entry whose information content may be changed from time to time, or to a group of functionally similar entires whose corresponding components are of different lengths.

WRITE—(1) To transfer information, usually from main storage, to an output device; (2) to record data in a register, location, or other storage device or medium.

XEROGRAPHY—A dry copying process involving the photo electric discharge of an electrostatically charged plate. The copy is made by tumbling a resinous powder over the plate, the remaining electrostatic charge discharged and the resin transferred to paper or offset printing master.

ZERO ADDRESS INSTRUCTION—An instruction consisting of an operation which does not require the designation of an address in the usual sense; e.g., the instruction, "shift left 0003," has in its normal address position the amount of the shift desired.

ZONE—(1) A portion of internal storage allocated for a particular function or purpose. (2) The three top positions of 12, 11, and 10 on certain punch cards. In these positions, a second punch can be inserted so that with punches in the remaining positions 1 to 9, alphabetic characters may be represented.

# Appendix II
# Commonly Used
# Abbreviations For Digital
# Electronics

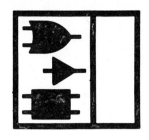

| | |
|---|---|
| ASCII | American Standard Code for Information Interchange |
| BCD | Binary Coded Decimal |
| BIT | Binary Digit |
| CAM | Content Addressable Memory |
| CML | Current Mode Logic |
| CMOS | Complementary MOS |
| DCTL | Direct Coupled Transistor Logic |
| DTL | Diode Transistor Logic |
| ECL | Emitter Coupled Logic |
| FET | Field Effect Transistor |
| FIFO | First In First Out |
| IC | Integrated Circuit |
| LED | Light Emitting Diode |
| LSB | Least Significant Bit |
| LSD | Least Significant Digit |
| LSI | Large Scale Integration |
| MNOS | Metal Oxide Nitride Semiconductor |
| MOS | Metal Oxide Semiconductor |
| MSB | Most Significant Bit |
| MSD | Most Significant Digit |
| MSI | Medium Scale Integration |
| MUX | Multiplex |
| RAM | Random Access Memory |
| RCTL | Resistor Capacitor Transistor Logic |
| RMM | Read Mostly Memory |
| ROM | Read Only Memory |
| RTL | Resistor Transistor Logic |
| TTL | Transistor Transistor Logic |
| $T_2L$ | Transistor Transistor Logic |

| $n$ | $n^4$ | $n^5$ | $n^6$ | $n^7$ | $n^8$ |
|---|---|---|---|---|---|
| 1 | 1 | 1 | 1 | 1 | 1 |
| 2 | 16 | 32 | 64 | 128 | 256 |
| 3 | 81 | 243 | 729 | 2187 | 6561 |
| 4 | 256 | 1024 | 4096 | 16384 | 65536 |
| 5 | 625 | 3125 | 15625 | 78125 | 390625 |
| 6 | 1296 | 7776 | 46656 | 279936 | 1679616 |
| 7 | 2401 | 16807 | 117649 | 823543 | 5764801 |
| 8 | 4096 | 32768 | 262144 | 2097152 | 16777216 |
| 9 | 6561 | 59049 | 531441 | 4782969 | 43046721 |
| | | | | | $\times 10^8$ |
| 10 | 10000 | 100000 | 1000000 | 10000000 | 1.000000 |
| 11 | 14641 | 161051 | 1771561 | 19487171 | 2.143589 |
| 12 | 20736 | 248832 | 2985984 | 35831808 | 4.299817 |
| 13 | 28561 | 371293 | 4826809 | 62748517 | 8.157307 |
| 14 | 38416 | 537824 | 7529536 | 105413504 | 14.757891 |
| 15 | 50625 | 759375 | 11390625 | 170859375 | 25.628906 |
| 16 | 65536 | 1048576 | 16777216 | 268435456 | 42.949673 |
| 17 | 83521 | 1419857 | 24137569 | 410338673 | 69.757574 |
| 18 | 104976 | 1889568 | 34012224 | 612220032 | 110.199606 |
| 19 | 130321 | 2476099 | 47045881 | 893871739 | 169.835630 |
| | | | | $\times 10^9$ | $\times 10^{10}$ |
| 20 | 160000 | 3200000 | 64000000 | 1.280000 | 2.560000 |
| 21 | 194481 | 4084101 | 85766121 | 1.801089 | 3.782286 |
| 22 | 234256 | 5153632 | 113379904 | 2.494358 | 5.487587 |
| 23 | 279841 | 6436343 | 148035889 | 3.404825 | 7.831099 |
| 24 | 331776 | 7962624 | 191102976 | 4.586471 | 11.007531 |
| 25 | 390625 | 9765625 | 244140625 | 6.103516 | 15.258789 |
| 26 | 456976 | 11881376 | 308915776 | 8.031810 | 20.882706 |
| 27 | 531441 | 14348907 | 387420489 | 10.460353 | 28.242954 |
| 28 | 614656 | 17210368 | 481890304 | 13.492929 | 37.780200 |
| 29 | 707281 | 20511149 | 594823321 | 17.249876 | 50.024641 |
| | | | $\times 10^8$ | $\times 10^{10}$ | $\times 10^{11}$ |
| 30 | 810000 | 24300000 | 7.290000 | 2.187000 | 6.561000 |
| 31 | 923521 | 28629151 | 8.875037 | 2.751261 | 8.528910 |
| 32 | 1048576 | 33554432 | 10.737418 | 3.435974 | 10.995116 |
| 33 | 1185921 | 39135393 | 12.914680 | 4.261844 | 14.064086 |
| 34 | 1336336 | 45435424 | 15.448044 | 5.252335 | 17.857939 |
| 35 | 1500625 | 52521875 | 18.382656 | 6.433930 | 22.518754 |
| 36 | 1679616 | 60466176 | 21.767823 | 7.836416 | 28.211099 |
| 37 | 1874161 | 69343957 | 25.657264 | 9.493188 | 35.124795 |
| 38 | 2085136 | 79235168 | 30.109364 | 11.441558 | 43.477921 |
| 39 | 2313441 | 90224199 | 35.187438 | 13.723101 | 53.520093 |
| | | | $\times 10^9$ | $\times 10^{10}$ | $\times 10^{12}$ |
| 40 | 2560000 | 102400000 | 4.096000 | 16.384000 | 6.553600 |
| 41 | 2825761 | 115856201 | 4.750104 | 19.475427 | 7.984925 |
| 42 | 3111696 | 130691232 | 5.489032 | 23.053933 | 9.682652 |
| 43 | 3418801 | 147008443 | 6.321363 | 27.181861 | 11.688200 |
| 44 | 3748096 | 164916224 | 7.256314 | 31.927781 | 14.048224 |
| 45 | 4100625 | 184528125 | 8.303766 | 37.366945 | 16.815125 |
| 46 | 4477456 | 205962976 | 9.474297 | 43.581766 | 20.047612 |
| 47 | 4879681 | 229345007 | 10.779215 | 50.662312 | 23.811287 |
| 48 | 5308416 | 254803968 | 12.230590 | 58.706834 | 28.179280 |
| 49 | 5764801 | 282475249 | 13.841287 | 67.822307 | 33.232931 |
| 50 | 6250000 | 312500000 | 15.625000 | 78.125000 | 39.062500 |

| $n$ | $n^4$ | $n^5$ | $n^6$ | $n^7$ | $n^8$ |
|---|---|---|---|---|---|
| | | | $\times 10^9$ | $\times 10^{11}$ | $\times 10^{13}$ |
| 50 | 6250000 | 312500000 | 15.625000 | 7.812500 | 3.906250 |
| 51 | 6765201 | 345025251 | 17.596288 | 8.974107 | 4.576794 |
| 52 | 7311616 | 380204032 | 19.770610 | 10.280717 | 5.345973 |
| 53 | 7890481 | 418195493 | 22.164361 | 11.747111 | 6.225969 |
| 54 | 8503056 | 459165024 | 24.794911 | 13.389252 | 7.230196 |
| 55 | 9150625 | 503284375 | 27.680641 | 15.224352 | 8.373394 |
| 56 | 9834096 | 550731776 | 30.840979 | 17.270948 | 9.671731 |
| 57 | 10556001 | 601692057 | 34.296447 | 19.548975 | 11.142916 |
| 58 | 11316496 | 656356768 | 38.068693 | 22.079842 | 12.806308 |
| 59 | 12117361 | 714924299 | 42.180534 | 24.886515 | 14.683044 |
| | | $\times 10^8$ | $\times 10^{10}$ | $\times 10^{11}$ | $\times 10^{13}$ |
| 60 | 12960000 | 7.776000 | 4.665600 | 27.993600 | 16.796160 |
| 61 | 13845841 | 8.445963 | 5.152037 | 31.427428 | 19.170731 |
| 62 | 14776336 | 9.161328 | 5.680024 | 35.216146 | 21.834011 |
| 63 | 15752961 | 9.924365 | 6.252350 | 39.389806 | 24.815578 |
| 64 | 16777216 | 10.737418 | 6.871948 | 43.980465 | 28.147498 |
| 65 | 17850625 | 11.602906 | 7.541889 | 49.022279 | 31.864481 |
| 66 | 18974736 | 12.523326 | 8.265395 | 54.551607 | 36.004061 |
| 67 | 20151121 | 13.501251 | 9.045838 | 60.607116 | 40.606768 |
| 68 | 21381376 | 14.539336 | 9.886748 | 67.229888 | 45.716324 |
| 69 | 22667121 | 15.640313 | 10.791816 | 74.463533 | 51.379837 |
| | | $\times 10^8$ | $\times 10^{10}$ | $\times 10^{12}$ | $\times 10^{14}$ |
| 70 | 24010000 | 16.807000 | 11.764900 | 8.235430 | 5.764801 |
| 71 | 25411681 | 18.042294 | 12.810028 | 9.095120 | 6.457535 |
| 72 | 26873856 | 19.349176 | 13.931407 | 10.030613 | 7.222041 |
| 73 | 28398241 | 20.730716 | 15.133423 | 11.047399 | 8.064601 |
| 74 | 29986576 | 22.190066 | 16.420649 | 12.151280 | 8.991947 |
| 75 | 31640625 | 23.730469 | 17.797852 | 13.348389 | 10.011292 |
| 76 | 33362176 | 25.355254 | 19.269993 | 14.645195 | 11.130348 |
| 77 | 35153041 | 27.067842 | 20.842238 | 16.048523 | 12.357363 |
| 78 | 37015056 | 28.871744 | 22.519960 | 17.565569 | 13.701144 |
| 79 | 38950081 | 30.770564 | 24.308746 | 19.203909 | 15.171088 |
| | | $\times 10^8$ | $\times 10^{10}$ | $\times 10^{12}$ | $\times 10^{14}$ |
| 80 | 40960000 | 32.768000 | 26.214400 | 20.971520 | 16.777216 |
| 81 | 43046721 | 34.867844 | 28.242954 | 22.876792 | 18.530202 |
| 82 | 45212176 | 37.073984 | 30.400667 | 24.928547 | 20.441409 |
| 83 | 47458321 | 39.390406 | 32.694037 | 27.136051 | 22.522922 |
| 84 | 49787136 | 41.821194 | 35.129803 | 29.509035 | 24.787589 |
| 85 | 52200625 | 44.370531 | 37.714952 | 32.057709 | 27.249053 |
| 86 | 54700816 | 47.042702 | 40.456724 | 34.792782 | 29.921793 |
| 87 | 57289761 | 49.842092 | 43.362620 | 37.725479 | 32.821167 |
| 88 | 59969536 | 52.773192 | 46.440409 | 40.867560 | 35.963452 |
| 89 | 62742241 | 55.840594 | 49.698129 | 44.231335 | 39.365888 |
| | | $\times 10^9$ | $\times 10^{11}$ | $\times 10^{13}$ | $\times 10^{15}$ |
| 90 | 65610000 | 5.904900 | 5.314410 | 4.782969 | 4.304672 |
| 91 | 68574961 | 6.240321 | 5.678693 | 5.167610 | 4.702525 |
| 92 | 71639296 | 6.590815 | 6.063550 | 5.578466 | 5.132189 |
| 93 | 74805201 | 6.956884 | 6.469902 | 6.017009 | 5.595818 |
| 94 | 78074896 | 7.339040 | 6.898698 | 6.484776 | 6.095689 |
| 95 | 81450625 | 7.737809 | 7.350919 | 6.983373 | 6.634204 |
| 96 | 84934656 | 8.153727 | 7.827578 | 7.514475 | 7.213896 |
| 97 | 88529281 | 8.587340 | 8.329720 | 8.079828 | 7.837434 |
| 98 | 92236816 | 9.039208 | 8.858424 | 8.681255 | 8.507630 |
| 99 | 96059601 | 9.509900 | 9.414801 | 9.320653 | 9.227447 |
| 100 | 100000000 | 10.000000 | 10.000000 | 10.000000 | 10.000000 |

# Index